Stress in the Workplace
Past, Present and Future

Edited by
JACK DUNHAM
Management and Stress Management Consultant, London

W

WHURR PUBLISHERS

LONDON AND PHILADELPHIA

© 2001 Whurr Publishers
First published 2001 by
Whurr Publishers Ltd
19b Compton Terrace, London N1 2UN, England and
325 Chestnut Street, Philadelphia PA 1906, USA

British Library Cataloguing in Publication Data
A catalogue record for this book is available from the
British Library.

ISBN: 1 86156 181 4

Printed and bound in the UK by Athenaeum Press Ltd,
Gateshead, Tyne & Wear

Contents

Cary Cooper:
An Appreciation

This book consists of nine chapters written by internationally known and respected authors and research workers to honour Cary Cooper's many distinguished and distinctive contributions to occupational stress. He carried out a substantial amount of early research in this area in the late 1970s which helped to identify the physical health risks associated with repetitive work and other stressors in the industrial and public sector work environments. He worked (with Lennart Levi) in a series of consultations and workshops organized by the World Health Organization (WHO), and he was an adviser for the International Labour Office (ILO). He developed a measurement approach for the identification of occupational stress indicators in 1988 with Stephen Sloan and Stephen Williams. He has continued to contribute to the field of occupational stress in a highly productive and successful way in the 1990s, producing the *Handbook of Stress, Medicine and Health* in 1996, the same year that he published *Teachers Under Pressure: Stress in the Teaching Profession* with Cheryl Travers. In 1998 his *Theories of Organisational Stress* was issued and in 1999, with Michiel Kompier as co-author, he published *Preventing Stress, Improving Productivity; European Case-Studies in the Workplace*.

He has made a contribution to the work and careers of most of the authors of *Stress in the Workplace: Past, Present and Future* as research associate, PhD supervisor and examiner, editorial guide and mentor and academic colleague. The contributors to one chapter of this volume (James Campbell Quick et al.) have described Cary Cooper as having 'a pivotal role in the field of occupational stress. It is an understatement to say he is extraordinarily prolific. However it is not his prolific nature which is his most distinguishing characteristic. Through our chapter and our own line of research which he helped to nurture, we suggest that Professor Cooper's most distinguishing characteristic is his self-reliance, as manifest through his generosity.'

Cary Cooper is Professor of Organizational Psychology at the Manchester School of Management and Pro-Vice Chancellor (External Activities) of the University of Manchester Institute of Science and Technology (UMIST). He was the Founding Editor-in-Chief of the *Journal of Organizational Behavior* and is co-editor of the medical journal *Stress Medicine*. Cary is a Fellow of the British Psychological Society, the Royal Society of Arts, the Royal Society of Medicine and the Royal Society of Health. He is the President of the British Academy of Management. He also finds time (how?) to be a frequent contributor to national newspapers, TV and radio.

Jack Dunham

Contributors

Dr Susan Cartwright is Senior Lecturer in Organizational Psychology at the Manchester School of Management, UMIST and Director of the part-time Doctoral Programme. She is also Editor of the *Leadership and Organisation Development Journal* and Book Review Editor for *Stress Medicine*. She has published widely in the area of mergers and acquisitions and occupational stress. Her recent publications (with Professor Cary L. Cooper) include *Managing Mergers and Acquisitions* (1996), published by Butterworth Heinemann, and *Managing Workplace Stress* (1996), published by Sage.

Marilyn Davidson is Professor of Managerial Psychology in the Manchester School of Management, University of Manchester Institute of Science and Technology. She is author/co-author of 14 books on women at work, women in management and occupational stress. Three of her most recent books are *Shattering the Glass Ceiling – The Woman Manager*, *Women in Management: Current Research Issues Volume II* and *The Black and Ethnic Minority Woman Manager – Cracking the Concrete Ceiling* (which was shortlisted for the Best Management Book of the Year). She has been employed as a research consultant for numerous organizations and government bodies, and she is a Chartered Psychologist and Fellow of the British Psychological Society and the Royal Society of Arts.

Sandra Fielden is Director of the Centre for Business Psychology, with interests in gender, diversity, equal opportunities, occupational stress, unemployment, small business success, psychological contract and organizational change. The Centre is involved in applied research and consultancy activities within the public and private sector, including European funded research into gender issues and economic growth. She is currently the editor of *Women in Management Review* and lectures in several areas of psychology, including Gender and Diversity, at Manchester School of Management, UMIST.

Howard Kahn lectures in the School of Management at Heriot-Watt University, Edinburgh. He has published over 50 articles and books on occupational stress and other aspects of organizational behaviour, and is Book Review Editor of the *Journal of Occupational and Organisational Psychology*. Previously he worked as a computer and business analyst with British Steel and Lloyd's (Underwriters) of London. He acts as a consultant to various organizations on issues relating to people at work.

Michiel A.J. Kompier is Professor of Work and Organizational Psychology at the University of Nijmegen in the Netherlands. He has published several books and over 100 book chapters and articles, mainly in the field of work, stress and health. In 1999, together with Professor Cary Cooper, he published the book *Preventing Stress, Improving Productivity: European Case Studies in the Workplace*.

Tage S. Kristensen is Senior Researcher at the National Institute of Occupational Medicine, Copenhagen, Denmark. He is chairman of the Danish Society of Psychological Medicine and vice-chairman of the Scientific Committee of Cardiology in Occupational Health under the International Committee of Occupational Health (ICOH). He has conducted research in the fields of psychological factors at work, absenteeism, burnout, cardiovascular disease and work, psychological intervention studies, prevention, epidemiological methods in psychological research and medical sociology.

Lennart Levi became Sweden's first Professor of Psychosocial Medicine at the Karolinska Institute in 1978. Forty years ago he founded the Karolinska's Department of Stress Research, which in 1973 was designated the first WHO centre in its field. Since its founding in 1980 and until 1995, he also directed the National Swedish Institute for Psychosocial Factors and Health (IPM). Before and after his retirement in 1995, he has been a key figure in the WHO's, ILO's and EU's activities in this field. His over 300 scientific papers and books in many languages have influenced scientists and decision makers in many organizations and countries.

Debra L. Nelson is CBA Associates Professor of Business Administration and Professor of Management at Oklahoma State University (OSU). Her research interests include organizational stress management, newcomer socialization, and management of technology. Dr Nelson's research has been published in numerous leading academic and professional journals. In addition, she is co-author of: *Stress and Challenge at the Top: The Paradox of the Successful Executive* (John Wiley & Sons, 1990) and

Organizational Behaviour: Foundations, Realities and Challenges, Third Edition (West Publishing, 2000). She was the recipient of the Regents' Distinguished Teaching Award in 1994 and the Burlington Northern Faculty Achievement Award at OSU in 1991. Dr Nelson also serves on the editorial review boards of the *Journal of Occupational Health and Psychology* and the *Journal of Organizational Behavior*.

Sheila Panchal graduated in Psychology at Leeds University and is a postgraduate researcher at Manchester School of Management.

James Campbell Quick is Professor of Organizational Behaviour at the University of Texas at Arlington and Associate Editor, Academy of Management Executive. Dr Quick is a Fellow of the Society for Industrial and Organizational Psychology, the American Psychological Association, and the American Institute of Stress. He was APA's stress expert to the National Academy of Sciences on National Health Objectives for the Year 2000. Dr Quick has been recognized by the American Heart Association with the Texan Volunteer Recognition Award and by the Colgate University Alumni Corporation with The Maroon. Colonel Quick is in the US Air Force Reserve and serves as Senior Individual Mobilization Augmentee at the San Antonio Air Logistics Centre (AFMC), Kelly AFB, Texas. His military awards and decorations include the Meritorious Service Medal and National Defense Service Medal with Bronze Star.

Jonathan D. Quick is Program Director for the World Health Organization in Geneva, Switzerland, and Adjunct Associate Professor of Public Health at Boston University. He has authored or edited six books and over 30 articles in stress management, preventative medicine and public health management. He was an officer in the US Public Health Service and has spent the last twelve years in international public health. He served as health services development adviser for the Afghan mujaheddin in Peshawar, Pakistan, and helped design a health financing system in East Africa while based in Nairobi, Kenya. A Fellow of the American College of Preventative Medicine, a Fellow of the Royal Society of Medicine (UK) and listed in the 1995/96 International Who's Who in Medicine, he led the way with his brother in transferring the public health concepts of prevention from preventative medicine to organizational stress.

Johannes Siegrist studied sociology, philosophy and history at the University of Freiberg, Germany. From 1973 to 1992 he was Professor of Medical Sociology at the University of Marburg. Since 1992 he has been Professor of Medical Sociology and Director of the Postgraduate Training

Program on Public Health at the Medical School, University of Dusseldorf, Germany. His major research activities are on psychosocial determinants of chronic disease, especially work stress and cardiovascular disease. In this framework he has developed and extensively tested the model of effort–reward imbalance. His additional research activities concern socio-logical evaluation research in medicine (e.g. patient/physician interaction, new models of health care) and health (e.g. social dimensions of quality of life). He is the author of some 200 original papers and the author or editor of some 15 books, including a standard German textbook on medical sociology. He has held a visiting professorship at the Johns Hopkins University School of Public Health, Baltimore, Maryland, USA, and at the Institute of Advanced Studies, Vienna, Austria. His honours include the Hans Roemer Award of the German College of Psychosomatic Medicine. He has held the Belle van Suylen Chair at the University of Utrecht, Netherlands, and he has an honorary membership of the European Society of Health and Medical Sociology.

Töres Theorell became a licensed physician in 1967. He served as a physi-cian in internal medicine for 11 years and then in social medicine, general practice and occupational medicine in Stockholm. In 1981 he became a professor of health care service at the National Institute for Psychosocial Factors. In 1995 he succeeded Lennart Levi as director of the Institute and also became a professor at the Karolinska Institute. His main writings have been in psychosocial factors and stress mechanisms, in particular in relation to myocardial infarction, hypertension and functional gastroin-testinal disorders. He has also conducted several intervention studies.

Cheryl Travers is a lecturer in Organizational Behaviour and Human Resource Management at Loughborough University Business School. She has trained a large number of teachers in stress management and is the author of a considerable number of publications on stress in teaching. She was the joint author with Cary Cooper of a large-scale investigation commissioned by the National Association of Schoolmasters and Union of Women Teachers (NASUWT).

Preface

Hindsight suggests 'A Global Perspective' should have been added to the title of this book. The first two chapters are directly concerned with global concerns. Lennart Levi reminds us that 'global' has two meanings: worldwide and comprehensive. He has presented a clear psychosocial framework for understanding sickness and health in the work and non-work aspects of people's lives world-wide. The second chapter, written by James Campbell Quick, Debra Nelson and James' brother Jonathan Quick, also has a global perspective. They argue that 'this is a time of dramatic global change and risk'. They present an account of their research with self-reliant executives in industry and the military (James Quick is a Colonel in the US Air Force Reserve) who are able to develop good interpersonal relationships and avoid social isolation – and be successful.

The next two chapters provide valuable insights into the meaning of these changes and the stress effects associated with the increased risks that accompany significant change. Töres Theorell has focused his research on the 'demand–control model' (the relationship between demands on workers and the control they have or do not have over their work). He reports evidence to show an increasing imbalance between demand and control leading to 'a lack of decision latitude' for many workers and an increased risk of cardiovascular disease. Johannes Siegrist is also concerned with an imbalance – this time between effort and reward at work (or at home for that matter). If this particular lack of equilibrium results in high cost/low gain conditions then, according to Siegrist's evidence, stress develops at the emotional and psychological levels.

Susan Cartwright and Sheila Panchal research the stress of mergers and acquisitions (M&As). They report that the incidence of mergers and acquisitions has increased significantly during the 1990s, and they suggest that it is unlikely that the frequency of M&As will diminish in the near future. This has important implications for the incidence of occupational stress because, the authors argue, M&As are 'large scale organizational change events' and the individuals who are affected by them 'are unlikely to have developed an effective repertoire of coping strategies to deal with them'.

Howard Kahn outlines research findings relating to the stress of working in the financial sector, as stockbrokers, money market traders, arbitrageurs, investment analysts and the like, for clearing banks, the London International Financial Futures Exchange, merchant banks and foreign owned banks.

The contribution by Sandra Fielden and Marilyn Davidson is on 'Stress and the Woman Manager'. It is only relatively recently that research into occupational stress and women in managerial positions has received any real attention. Their research evidence clearly demonstrates that, in comparison to their male counterparts, female managers are faced with many additional sources of stress arising as a direct result of discrimination and prejudice in the workplace and increased home/work conflicts.

Cheryl Travers' analysis of the causes and the rising costs of teacher stress is very relevant for policy makers and managers. Her emphasis is on organizational factors and she makes a strong appeal for urgent, relevant research.

The final chapter is concerned with intervention. Michiel Kompier and Tage Kristensen, in their review of research into stress prevention strategies, are critical of their mainly individual orientation and lack of adequate evaluation. They have provided a framework of guidance for present and future research workers and a rigorous academic resource for professionals involved in further study for higher degrees.

I have enjoyed working with these distinguished contributors and I thank them most warmly for their enthusiastic support for this project to honour Cary Cooper. It has been an exceptional learning experience for me to be the co-ordinator of their presentations on workplace stress; past present and future.

Jack Dunham
August 2000

Chapter 1
Stress in the Global Environment

LENNART LEVI

Introduction

This chapter is about stress, its causes, manifestations and conse-quences, in the *global* environment. 'Global' means 'of, relating to, or involving the entire earth', but also 'comprehensive, total'. An attempt will be made to combine these two approaches. It will also discuss some options for the prevention of stress and stress-related ill health, and for health promotion. Needless to say, all this will be made from a birds-eye view.

It goes without saying that the *biomedical* approach in basic and clinical research concerning such issues and the application of its results in health action are of the utmost importance. It is equally evident that it is far from sufficient for that purpose. The awareness that this is so is the basis of the expression, 'the other half of medicine', which refers to the complementary *psychosocial* approach to stress-related ill health, health promotion and health care.

Another complementary approach is summarized in the concept of '*investment* for health'. The verb 'invest' can be defined as a 'commitment (of money or other resources) in order to gain a financial return, to spend or devote for future advantage or benefit'. Consequently, an investment for health refers to a commitment of resources in order to gain a return that is above and beyond the size of the investment. Seen in such a way, the investment does not constitute a burden, but an opportunity for increasing returns.

This paper is an elaboration and update of Levi L (1998) The other half of medicine: the concept of psychosocial stressors, and its implications for health and the health profes-sions. Forum Trends in Experimental and Clinical Medicine 8(3): 36–46, suppl. 4.

1

However, given the complexity of the problems under discussion, such 'investments' need to be based on a *systems approach*. A 'system' can be seen as 'a group of interacting, interrelated, or interdependent elements forming a complex whole'. The term is derived from late Latin 'systema', from Greek 'sustema' – to combine. A systems analysis accordingly implies 'the study of an activity or a procedure to determine the desired end and the most efficient method(s) of obtaining this end'. (American Heritage College Dictionary, 1993).

One of the few stress researchers who understood and actively contributed to this way of thinking is Cary L. Cooper. We worked together throughout the 1980s in a series of consultations and workshops organized by the World Health Organization (WHO), the International Labour Office (ILO), or both (Kalimo, El-Batawi and Cooper, 1987).

Uni- or multi-causal aetiology

German pathologist Rudolf Virchow (1821–1902) formulated a die-hard dogma stating that every disease is conditioned by a single specific and clearly defined cause – for example, the tubercle bacillus in the case of tuberculosis. The 'father' of the biological stress concept, the Canadian biologist Hans Selye (1907–1982), in turn, demonstrated that exposure to *a wide variety* of stressors and their interaction could lead to similar patterns of changes in the functioning and structure of organs and organ systems. His findings laid the foundation for today's more complex ecological view of aetiology, pathogenesis and salutogenesis. This view is based on the awareness that every organ and system of organs can be affected, both by psychosocial and physicochemical stressors and by corresponding salutogenic stimuli. The type and degree of the outcomes of such an influence depend on the type, intensity and duration of stimuli, the 'programming' of the organism and the conditions under which the exposure occurs. The *negative* effects that need to be identified and prevented are due to components in the person–environment ecosystem that are necessary, sufficient or contributory, in:

1. causing disease;
2. accelerating its course;
3. triggering its symptoms.

It is, of course, at least equally important to identify and promote *positive* system components that are necessary, sufficient or contributory in causing or promoting health and well-being.

Common stressors in the global environment

The notion that *physical* and *chemical* environmental influences can damage health and well-being is widely established and accepted – for example in the case of bacteria, viruses, nuclear radiation, short asbestos fibres, lead, mercury and organic solvents. However, it is harder to demonstrate and to find acceptance for the notion that *psychosocial* influences brought about by social conditions and conveyed by processes within the central nervous system and/or human behaviours can have corresponding effects (Levi and Andersson, 1974; Levi, 1979; Kompier and Levi, 1994).

Some of the *root causes* of both types of noxious influences have been described recently by the United Nations (UN, 1995) and United Nations Development Programme (UNDP, 1997). According to these analyses, the world's most impoverished people lack much more than a decent income. They are usually malnourished, badly housed, barely literate, easy prey for disease, and, more often than not, powerless to change their situation.

It is estimated that 1.3 billion people – more than one of every five on earth – live in debilitating poverty; more than a billion people around the world struggle to survive on less than a dollar a day.

The fact that extreme poverty and its multitude of stressors persist throughout the world despite the enormous generation of wealth over the past 50 years is one of the central reasons that development policies are under particular scrutiny.

Poverty has different faces. It can exist as endemic mass poverty in poor countries, pockets of poverty amidst wealth in rich countries, sudden poverty resulting from disasters and conflicts, temporary poverty from loss of employment, or the marginal poverty of those performing menial but essential work for inadequate wages. Behind these faces of poverty lies the grim reality of desperate lives without choices and, often, governments that lack the capacity to cope. There are also serious social and health problems in countries with economies in transition and countries experiencing fundamental political, economic and social transformations (Cornia, 1994). Within many societies, in both developed and developing countries, the gap between rich and poor has increased. Furthermore, despite the fact that the economies of some developing countries are growing rapidly, the gap between developed and many developing countries, particularly the least developed countries, has widened (UNDP, 1997). It would, indeed, be highly surprising if social stressors such as these did *not* have a major impact on the morbidity and mortality of hundreds of millions of people world-wide, not to mention the impact on their well-being and quality of life.

Fortunately, more and more health professionals and policy makers are beginning to realize that we cannot look to biomedicine alone to find the solutions to these problems. We need to look further afield and also take into consideration the socio-economic and psychosocial conditions of life, the ways people cope with these conditions and the resulting stress and health effects. These constitute the combination of discomfort and the organism's maladaptive 'step into overdrive', or 'giving-up' reaction patterns, which can damage health and well-being.

Determinants of health

Present knowledge of the main determinants of health comes from epidemiological as well as experimental research, in animals as well as in humans.

A recent in-depth review of this knowledge can be found in a special issue of *Acta Physiologica Scandinavica* (Folkow, Schmidt and Urhäs-Moberg, 1997) compiled and published in memory of James P. Henry and in a review paper by McEwen (1998).

A related approach has been chosen by Wilkinson (1996), who presents a strong case for his statement that social, rather than material, factors are now the limiting component in health and quality of life in developed societies. Poorer people in such countries may have annual death rates anywhere between twice and four times as high as richer people in the same society. However, this is not simply an effect of the absolute level of income but is more related to what he refers to as the effects of social relativities. In the developed world, it is not the richest countries that have the best health but the most egalitarian, namely those with a relatively small income gap between the richest and the poorest, those characterized by social cohesion, social morality and social capital.

Wilkinson draws attention to the fact that blue-collar workers almost invariably exhibit a much higher morbidity and mortality than white-collar workers. This is true for every major group of diseases – infections, cancers, cardiovascular, nutritional and metabolic diseases, respiratory diseases, accidents, and nervous and mental illnesses. All of them show a social class gradient. This is also true for 'all-cause' mortality, also after controlling for the effects of major individual risk factors. He points to 'the toxicity of social circumstances and patterns of organisation' and demonstrates their effects on health as mediated by either the psychosocial factors–stress–health linkage and/or by the psychosocial factors–health related behaviour–health linkage.

Summarizing available evidence, Wilkinson and Marmot (1998) draw attention to the fact that – even in the richest countries – the better off live several years longer and have fewer illnesses than the poor. 'These differences in health are an important social injustice, and reflect some of the most powerful influences on health in the modern world. People's lifestyles and the conditions in which they live and work strongly influence their health and longevity.' (Wilkinson and Marmot, 1998, p. 6.)

> Medical care can prolong survival after some serious diseases, but the social and economic conditions that affect whether people become ill are more important for health gains in the population as a whole. Poor conditions lead to poorer health. An unhealthy material environment and unhealthy behaviour have direct harmful effects, but the worries and insecurities of daily life and the lack of supportive environments also have an influence. (Wilkinson and Marmot, 1998, p. 7)

This insight was expressed about two decades ago by the US Surgeon-General (Califano, 1979) in three summarizing sentences about what he referred to as the 'modern killers':

- we are killing ourselves by our own careless habits;
- we are killing ourselves by carelessly polluting the environment; and
- we are killing ourselves by permitting harmful social conditions to persist.

All three statements relate, one way or another, to human *behaviours*. All these behaviours can be influenced, both therapeutically and preventively.

It is interesting to note that the awareness of this was, again, manifested in the mid-1980s, and not only by a few scientists or practitioners, but by the Swedish Government.

Following one decade of review and scientific consultations, this awareness was summarized in the Swedish Government's 'Public Health Service (Lines of Development) Bill', No. 1984/85:181, approved by the Parliament of Sweden on 21 March 21 1985. Some excerpts:

> Our health is determined in large measure by our living conditions and lifestyle.

> The health risks in contemporary society take the form of, for instance, work, traffic and living environments that are physically and socially deficient, unemployment and the threat of unemployment; abuse of alcohol and drugs, consumption of tobacco, unsuitable dietary habits, as well as psychological and social strains associated with our relationships – and lack of relationships – with our fellow beings.

> These health risks ... are now a major determinant of our possibilities of living a healthy life. This is true of practically all the health risks which give rise to

today's most common diseases, e.g. cardiovascular disorders, mental ill health, tumours and allergies, as well as accidents.

(Therefore), care must start from a holistic approach ... By a holistic approach we mean that people's symptoms and illnesses, their causes and consequences, are appraised in both a medical and a psychological and social perspective.

Instead of changing, however, such man-made conditions governing our lives continue to prevail. They are changing in a freewheeling manner and at an ever increasing rate, without taking into account our biological heredity, the boundaries that this sets on our ability to adapt, and the health outcomes (Harrison and Ziglio, 1998). This has subsequently led to a rather dramatic change in the overall spectrum of diseases. Neither social policy nor medical or other health-related research and training have taken these rapid changes into consideration and adapted to them.

Some obstacles

It seems likely that the present state of affairs is also due to the fragmentation both of research, higher education and training, and of administration and policy making. There is an emphasis in these activities on specific, single-disciplinary sectors. There is a reluctance to integrate approaches from relevant disciplines as well as sectors. This is so despite growing awareness that both the health and the health care of the population is conditioned by factors that act across both disciplines and societal sectors. Examples of such health and health care problems relate to (1) suicide and suicide attempts, (2) ischaemic heart disease, and (3) several malignant diseases. Although the biomedical approach is a *sine qua non* for their study, treatment and prevention, there is no doubt that socio-economic and psychosocial approaches are badly needed but so far usually lacking.

One major explanation for this state of affairs is the extreme complexity of the systems under consideration. Ideally, diagnosis, therapy and prevention should take this complexity into account by applying a systems approach. Instead, both political and professional approaches are usually simplistic – either as fragmented, ad hoc actions, or, worse, as a search for, or belief in, a 'silver bullet' that is expected to solve *all* problems, in all respects.

Definitions; a theoretical model

According to James Grier Miller (personal communication), 'one should simplify as much as possible – *but not more*'. An attempt to do this is presented by Kagan and Levi (1975), and later with minor modifications by Levi (1997). But first we need to define our terms (Kagan and Levi, 1975; Levi, 1979).

Psychosocial factors arise from social arrangements, are mediated through perception and experience (higher nervous processes) and include:

- structures and processes in the total environment that can elicit pathogenic (or, conversely, salutogenic, i.e. health-promoting) effects (e.g. personality, customs, attitudes);
- psychosocially induced emotional, cognitive, behavioural and physiological mechanisms leading to disease (e.g. anxiety, depression; difficulties with memory or concentration; abuse of alcohol or tobacco; overproduction of various hormones); or promoting health and well-being;
- psychosocially induced mental and physical disease; and decrease in well-being, satisfaction, and quality of life;
- aspects of the health care system, including health promoting measures, and also including influences on the effectiveness and efficiency of health care.

Starting from such a perspective, common denominators in the causation of psychosocially induced ill health are: the discrepancy between human needs and environmental possibilities for their satisfaction; the discrepancy between human capacity and environmental demands; and the discrepancy between human expectations and the situation as perceived (Figure 1.1, boxes 2 and 3). Such discrepancies are common in times of environmental deprivation or excess, when there is conflict between social roles, or when social change is rapid and there are no generally accepted rules of conduct (Cornia, 1994).

According to this model, Kagan and Levi (1975), we are surrounded by nature (Figure 1.1, box 1), whose influences on us we modify and adjust by social arrangements, i.e. social structures and processes (box 2). These influence us through our senses. Their actions are experienced, filtered and appraised by the brain, sometimes resulting in psychosocial stimuli (box 3). These act on a human organism characterized by a psycho-biological programme (box 4), conditioned by earlier environmental influences and genetic factors. Some of the interactions between all these factors make the organism react. Some of these reactions are related to health, while others are not. In this context we focus on the former. Some of these mechanisms (box 5) are *specific* in the sense that they are related to one individual stressor or to certain individual characteristics of the organism, or lead to a specific type of morbidity or mortality. Others are *non-specific* in the sense that they are triggered by many conditions, in many types of individuals and

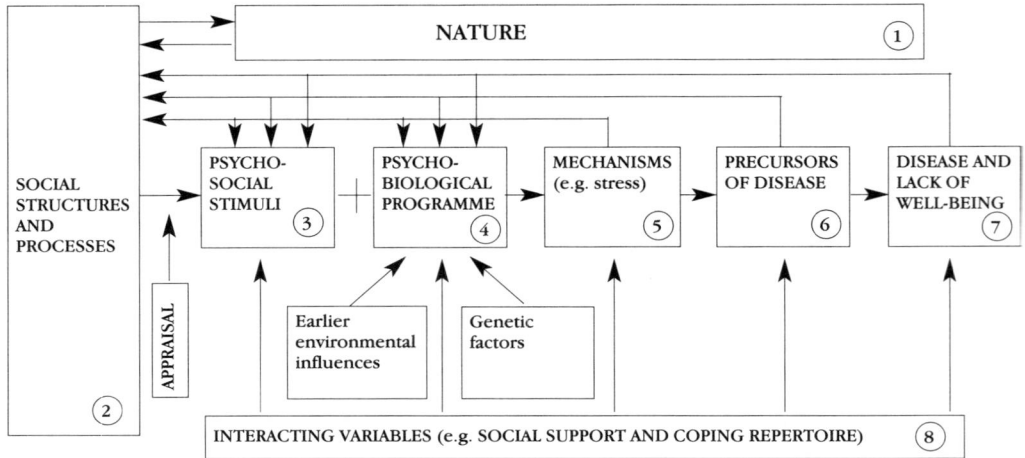

Figure 1.1: Human ecological system. Human element detailed (Adapted from Kagan and Levi, 1975.)

relate to many types of morbidity and mortality. The latter have been defined as *stress* (Selye, 1936). These mechanisms might lead to precursors of disease (box 6) or to disease and lack of well-being (box 7). This sequence of events is not a one-way flow but takes place in a system with feedback loops. What occurs in boxes 5 through 7 acts back on the social structures and processes, their appraisal, the resulting stimuli and the psycho-biological programme, sometimes creating vicious circles. This flow of events is modified by interacting variables (box 8), the most important ones being the presence or absence of *social support* and its utilization, and the *coping repertoire* of the individual in terms of problem or emotion oriented approaches.

What we need to identify is:

- the content of each box;
- the interaction between any of the boxes; and
- the dynamics of the entire system.

Options for disease prevention and health promotion

Once this has been achieved, we can try to prevent disease and promote health by addressing:

- social structures and processes;

- the way people appraise these social structures and processes;
- the resulting stimuli;
- the psycho-biological programme;
- the pathogenic emotional, cognitive, behavioural and physiological mechanisms;
- the precursors of disease;
- the diseases and the lack of well-being; and
- the interacting variables (by improving social support and coping repertoire).

Of course, to become more effective, these approaches could and should be combined and integrated.

The above description obviously has a bias towards interventions against aetiology and pathogenesis. However, it could, and should, be complemented by a corresponding *promotion of salutogenesis*.

Needless to say, not all (or even most) psychosocial (or physical, or chemical) stimuli act pathogenetically. Some have no effects on health, whilst others do counteract disease, or even promote health and well-being. In medicine, the emphasis is, and has always been, on *negative* outcomes and what may lead to them – on pathogenesis, morbidity and mortality, i.e. on *pathology*. The latter is the scientific study of the nature of disease and its causes. It comes from the prefix *patho* from Latin and Greek *pathos*, suffering. *Genesis*, the origin, the coming into being of something, comes from Latin and Greek. Accordingly, pathogenesis may be defined as 'the development of a diseased condition'. In contrast, something can be salutary, i.e. favourable to health, wholesome. The term is derived from old French *salutaire* from Latin *salutaris*, from *salus*, health. Analogously, salutogenesis (Antonovsky, 1987) could be defined as 'the development of a condition of health'.

According to the founding fathers of WHO (1946), health could be characterized as 'not only the absence of disease or infirmity but also a state of complete physical, mental and social well-being'.

How, then, can disease be prevented (and health promoted)? Theoretically, this can be done in accordance with principles spelt out in the EU Framework Directive (89/391/EEC), according to which employers have a:

> ... duty to ensure the safety and health of workers in every aspect related to the work, on the basis of the following *general principles of prevention*:

- avoiding risks;
- evaluating the risks which cannot be avoided;

- combating the risks at source;
- adapting the work to the individual, especially as regards the design of workplaces, the choices of work equipment and the choice of working and production methods, with a view, in particular, to alleviating monotonous work and work at a predetermined work rate and to reducing their effects on health;
- *developing a coherent overall prevention policy* [my italics] which covers technology, organization of work, working conditions, social relationships and the influence of factors related to the working environment. (89/391/EE)

Related approaches to disease prevention and investment for health should, of course, be considered also for non-work aspects of human life, including *level of living* areas such as:

- education and training;
- economic resources;
- housing;
- transport and communication;
- leisure and recreation;
- social relations;
- political resources;
- safety and security;
- health and medical services;
- equality and equity.

The EU Framework Directive restricts itself to conditions of work, thereby limiting itself to an '8-hours-a-day approach'. The remaining 16 hours should, of course, *also* be considered.

A systems approach

As mentioned above, to be effective, this requires a *systems approach*. A most interesting initiative along these lines has been made recently by the British Government, in its Green Paper 'Our Healthier Nation – A Contract For Health' (1998) and its subsequent White Paper 'Saving Lives: Our Healthier Nation' (1999). In essence, these 'papers' spell out five types of factors affecting health. The first category is referred to as 'fixed'. It includes the genes, sex and ageing of each individual, and is accordingly difficult to influence in a disease preventing and/or health promoting manner. In contrast, the other four categories could and should be approached:

- *social and economic* (such as employment, poverty, social exclusion);
- *environment* (such as air and water quality, housing and social environment);
- *lifestyle* (such as physical activity, diet, smoking, alcohol, sexual behaviour, drugs);
- *access to services* (such as education, health and social services, transport, and leisure).

All these and related factors can be dealt with in a co-ordinated, systems approach, *across societal sectors and levels*, in a Contract for Health. The three groups of *partners* in such a contract are:

- central government and national players;
- local players and communities;
- all citizens.

Examples of what these three categories of players can do are given in Table 1.1.

Briefly, then, this means that all stakeholders should act and collaborate:

- across societal sectors (health, social affairs, labour, education, housing, communication, etc.), and
- across societal levels (UN, EU, national, regional, local and 'grass-roots' levels).

As pointed out by Kickbusch (1997), the Investment for Health approach has far-reaching implications. Quoting the Human Development Report (UNDP, 1997) she points out that eradicating world poverty would cost only 1% of global income, and no more than 2-3% of the respective national incomes, and that this investment would also eliminate a significant part of the global disease burden.

It is likely that the benefits, not only in human but also in economic terms, would far exceed the necessary investment. As support for such a view, she quotes a report by the World Bank (1997), according to which:

- global health spending amounted to US$2330 billion, i.e. 9% of global GDP (1994);
- US$250 billion, i.e. 11% of this, concerned middle and low-income countries;
- 84% of the world's population live in these countries;

Table 1.1: A Contract for Health (British Government, 1998)

Government and national players can:	Local players and communities can:	People can:
Provide national co-ordination and leadership. Ensure that policy-making across Government takes full account of health and is well informed by research and the best expertise available. Work with other countries for international co-operation to improve health. Assess risks and communicate those risks clearly to the public. Ensure that the public and others have the information they need to improve their health. Regulate and legislate where necessary. Tackle the root causes of ill health.	Provide leadership for local health strategies by developing and implementing health improvement programmes. Work in partnerships to Improve the health of local people and tackle the root causes of ill health. Plan and provide high quality services to everyone who needs them.	Take responsibility for their own health and make healthier choices about their lifestyle. Ensure their own actions do not harm the health of others. Take opportunities to better their lives and their families' lives, through education, training and employment.

- they carry 93% of the world's disease burden;
- their health expenditure is expected to increase by US$9 billion annually;
- the latter would suffice to cover the preventive and curative services for the 900 million of the world's poor who still lack such services.

Kickbusch (1997) further draws attention to and makes a distinction between 'traditional hazards', related to poverty and insufficient develop-ment, and 'modern hazards', related to rapid development that lacks safeguards, and to unsustainable consumption. She considers health promotion to be 'a theory-based process of social change contributing to the goal of human development, building on many disciplines and applying interdisciplinary knowledge in a professional, methodical and creative way'. In her view, 'health promotion outcome' can be determined by an organized, partnership-based community effort contributing to health, quality of life and social capital of a society.

Thus, there is a growing awareness of the problems that people are experiencing both with regard to their social situation and to their health, well-being and quality of life. There is also increasing awareness of ways to prevent ill health and promote health and well-being. However, there still seems to be a long way to go before effective measures are taken to deal with existing problems, to prevent others from occurring, and to promote positive health. There is a wide *science–policy gap*.

A science–policy gap

This gap needs to be narrowed as much as possible by:

*   political decisions at all levels;
*   collaboration across societal sectors and levels, based on such decisions and policy statements;
*   education, training and information provided to all stakeholders, including those outside the health sector and those at the grass-roots level;
*   evaluation of the resulting policies, to make future actions more evidence-based.

An important provision for this at the EU level is the *Treaty of Amsterdam* (Article 152), according to which 'a high level of human health protection shall be ensured in the definition and implementation of *all* [my italics] Community policies and activities'.

If such a 'protection' is, indeed, 'ensured', preferably at *all* levels of government, *and* across all relevant societal sectors, one necessary (but not sufficient) prerequisite has been established for the prevention of stress-related morbidity and for the promotion of health and well-being.

A complementary, and equally necessary (but not sufficient) approach comprises corresponding action by all workers and citizens, i.e. empowered 'grass-roots' (British Government, 1998).

Skills for life

However, many hundreds of millions of people, and particularly the underprivileged ones, are so downhearted, helpless and have such low self-confidence that they give up, feel defeated, do not even start looking for new solutions. Some may have been taught in the past that the 'nanny state' would continue to provide for them, no matter what. Increasingly, no such service is given any longer, whilst many citizens have lost, or never developed, an ability to help themselves ('learnt helplessness' as described by Seligman (1975)). Others have never had access to any 'nanny'.

The WHO has attracted attention to this problem, based on a broad health promotion perspective. One of the WHO's ideas is to improve school-age children's 'introduction to life' by teaching them to live in a way that promotes health and well-being. Several hundred schools in Europe have been designated as 'healthy schools' and have been helped to educate the pupils to take care of their own health, as a complement to their ordinary curricula.

The 'healthy schools' complementary curriculum includes increasing the pupils' knowledge and understanding of a number of lifestyles known to be hazardous – smoking, drinking, illicit drugs, unhealthy diets, lack of exercise etc., in order to promote healthy lifestyles and avoid unhealthy ones.

In addition, the programme attempts to promote 'social skills' or 'skills for life' (WHO, 1994). The pupils are taught, for example, how to:

- communicate effectively;
- make decisions;
- solve problems;
- think critically;
- hold their own;
- resist peer pressure;
- manage their own worry, depression and stress;
- adapt to new environmental demands;
- get to know themselves.

Anybody possessing such 'skills for life' – related to but not identical with *emotional intelligence* (Stone and Dillehunt, 1978; Grant Consortium 1992; Goleman, 1995), will not remain unemployed for long. Nor will he or she remain in an unsatisfactory job. He or she will improve their job or find another.

The notion of 'skills for life' is very similar to the notion of 'everyday power' which was introduced by the Swedish Social Democratic Party (1996). As the term implies, it means having power and influence over one's own everyday life. This can be gained partly through life skills and partly by society not hindering individuals in applying them, as well as by individual and co-operative *bottom-up* efforts to solve problems. It is also assisted by promoting and encouraging such efforts, as a complement to society's own central and regional *top-down* resolution of problems. In this way we could get popular self-help movements against, for example, unemployment and other large-scale social and health problems.

Interesting examples of such 'empowerment' and attempts to create such 'learnt resourcefulness' and 'social capital' are reviewed and discussed by Putnam (1993), with a focus on co-operative approaches.

Social support

As pointed out by Corneil (1998), during the mid-1970s, public health practitioners, and in particular, epidemiologists introduced the concept of social support in their studies of causal relationships between stress, mortality and morbidity (Cassel, 1976; Cobb, 1976). Following up on these seminal works, investigators have moved away from considering social support as a unitary concept, and have attempted to understand the components of social stress, social support and their relation to health.

Hirsh (1980) described five possible elements of social support:

- *emotional support*: care, comfort, love, affection, sympathy;
- *encouragement*: praise, compliments; the extent to which one feels inspired by the supporter to feel courage, hope or to prevail;
- *advice*: useful information to solve problems; the extent to which one feels informed;
- *companionship*: time spent with an involved supporter; the extent to which one does not feel alone;
- *tangible aid*: practical resources, such as money or aid with chores;

A person who has access to all of this and can take advantage of it is likely to feel better and will also become more resistant to life's various trials (Johnson, 1986; Wilkinson, 1996; Corneil, 1998). Welfare will be improved along with health. It is, in fact, another important option for 'investment for health'.

Sense of coherence

When you are 'navigating on the ocean of life' it is good to have nautical charts and a compass with you. To have an idea of where you are heading and how and why. To have a salutogenic 'sense of coherence' (Antonovsky, 1987). The latter consists of three components:

- *Understandability*. People need to understand what is happening to them. Why aren't I getting a new job? Does the employer dislike me? Is there something I have done? Is it due to the indifference of the trade union? Or to the recession in the entire geographical region?
- *Manageability*. People need to be able to manage their current situation, to cope with it.
- *Meaningfulness*. People need to find a meaning in their present situations.

All of this can be taught to people in various problem scenarios, or to all people as a step towards improving their options for coping and their coping repertoire (Lazarus and Folkman, 1984).

The Verona Benchmark

An attempt to bring most of these strategies together is put forward by the World Health Organization in its Verona Benchmark (WHO, 1999). Its core principles are:

- focus on health, whatever the activity;
- full public participation;
- genuine intersectoral working (all societal sectors and levels);
- equity between and within populations, and between countries;
- sustainability (durable, resilient);
- a broad knowledge base (data, judgement, insight).

To promote this, all relevant variables should be *monitored*, with self-correcting loops for improved future policy-making. *Accountability* at all relevant levels should be secured. This applies both to stress prevention and management as well as to disease prevention and health promotion in general.

References

Antonovsky A (1987) Unravelling the Mystery of Health: How People Manage Stress and Stay Well. San Francisco, CA: Jossey-Bass.
British Government (1998) Our Healthier Nation. A Contract for Health. Green Paper. London: HMSO.
British Government (1999) Saving Lives. Our Healthier Nation. White Paper. London: HMSO.
Califano Jr, JA (1979) The Secretary's Foreword. In: Healthy People. The Surgeon General's Report on Health Promotion and Disease Prevention. US Department of Health, Education, and Welfare. Washington DC: Government Printing Office.

Cassel J (1976) The contribution of the social environment to host resistance. American Journal of Epidemiology 104: 107–23.

Cobb S (1976) Social support as a mediator of life stress. Psychosocial Medicine 38: 300–314.

Corneil DW (1998) Social Support. In: Stellman JM (ed.) Encyclopaedia of Occupational Health and Safety, vol. II, section 34.47. Geneva: International Labour Office.

Cornia GA (1994) Crisis in Mortality, Health and Nutrition. Economies in Transition Studies, Regional Monitoring Report No. 2. Florence: UNICEF.

Folkow B, Schmidt H, Urhäs-Moberg P (eds) (1997) Stress, health and the social environment. Acta Physiologica Scandinavica 161 (suppl.): 640.

Goleman D (1995) Emotional Intelligence. New York: Bantam.

Grant Consortium (1992) School-Based Promotion of Social Competence. In: Hawkins JD (ed.) Communities that Care. San Francisco, CA: Jossey-Bass.

Harrison D, Ziglio E (eds) (1998) Social determinants of health: implications for the health professions. Forum Trends in Experimental and Clinical Medicine 8(suppl. 4): 3.

Hirsh BJ (1980) Natural support systems and coping with major life changes. American Journal of Community Psychology 8: 159–71.

Johnson JV (1986) The Impact of Workplace Social Support, Job Demands and Work Control Upon Cardiovascular Disease in Sweden. PhD Dissertation, Baltimore, MD: Johns Hopkins University.

Kagan AR, Levi L (1975) Health and environment – psychosocial stimuli. A review. In: Levi L (ed.) Society, Stress and Disease – Childhood and Adolescence, vol. II. Oxford: Oxford University Press.

Kalimo R, El-Batawi M, Cooper CL (eds) (1987) Psychosocial Factors at Work and Their Relation to Health. Geneva: World Health Organization.

Kickbusch I (1997) Think Health. What Makes the Difference? Key Speech at the 4th International Conference on Health Promotion. Geneva: World Health Organization, WHO/HPR/HEP/41CHP/SP/97.1.

Kompier M, Levi L (1994) Stress At Work: Causes, Effects, and Prevention. A Guide

Lazarus RS, Folkman S (1984) Stress, Appraisal, and Coping. New York: Springer.

Levi L (1979) Psychosocial factors in preventive medicine. In: Hamburg DA, Nightingale EO, Kalmar V (eds): Healthy People. The Surgeon General's Report on Health Promotion and Disease Prevention. Background Papers. Washington DC: Government Printing Office.

Levi L (1997) Psychosocial environmental factors and psychosocially mediated effects of physical environmental factors. Scandinavian Journal of Work and Environmental Health 23 (suppl. 3):47–52.

Levi L, Andersson L (1974) Population, Environment and Quality of Life. A Contribution to the United Nations' World Population Conference. Stockholm: Royal Swedish Ministry of Foreign Affairs.

McEwen BS (1998) Protective and damaging effects of stress mediators. New England Journal of Medicine 338(3): 171–9.

Putnam RD (1993) Making Democracy Work. Civic Traditions in Modern Italy. Princeton, NJ: Princeton University Press.

Seligman MEP (1975) Helplessness. San Francisco, CA: WH Freeman.

Selye H (1936) A syndrome produced by diverse noxious agents. Nature 138: 32.

Stone KF, Dillehunt HQ (1978) Self Science. The Subject is Me. Santa Monica, CA: Goodyear.

Swedish Social Democratic Party (1996) Sweden Facing 2000. Guidelines Adopted at the Social Democratic Congress 15–17 March (Sverige inför 2000-talet. Riktlinjer antagna på socialdemokraternas kongress 15-17 mars 1996). Stockholm: Socialdemokraterna.

UN (1995) World Summit for Social Development. The Copenhagen Declaration and Programme of Action. New York: United Nations.

UNDP (1997) Human Development Report. New York: Oxford University Press.

WHO (1946) Preparatory Committee of the International Health Conference.

WHO (1994) Life Skills Education in Schools. World Health Organization, Division of Mental Health, Geneva (WHO/MNH/PSF/93.7A.Rev.1).

WHO (1999) The Verona Initiative. Investing for Health: The Economic, Social and Human Environment. Arena Meeting II. Copenhagen: WHO.

Wilkinson RG (1996) Unhealthy Societies. The Affliction of Inequality. London and New York: Routledge.

Wilkinson R, Marmot M (1998) The Solid Facts. Copenhagen: WHO.

World Bank (1997) Health, Nutrition and Population. Sector Strategy Study. Washington DC: World Bank.

Chapter 2
Occupational Stress and Self-reliance: Developmental and Research Issues

JAMES CAMPBELL QUICK, DEBRA L. NELSON AND
JONATHAN D. QUICK

Occupational stress was a major industrial concern of the 1990s because the 1990s have proved to be a time of dramatic global change in the world's industrial and economic systems. These global changes have resulted in significant human impacts, with threat effects for individuals and clear, measurable stress effects. Levinson (1996) has described this as the new age of self-reliance.

In this chapter the first section discusses the dimensions of global change in the new organizational reality, along with their human impact. The next section focuses on the paradox of self-reliance. The third section discusses the implication of self-reliance for social support, mentoring, and health. The final section considers research issues in occupational stress and self-reliance.

The new organizational reality

A Time of global change

This is a time of dramatic global change. A new organizational reality is emerging on the industrial landscape and it is being shaped by a set of economic, technological and competitive forces that span international boundaries (Gowing, Kraft and Quick, 1998). Economically, the boundaries between markets and around individual organizations are becoming increasingly permeable such that the next step in the evolution of organizations is the virtual corporation (see Martin and Freeman, 1998). Further, political and social changes are intertwined with economic change. Because organizations are centrally concerned with the management of economic risks, political and social change creates both uncertainty and risk as well as the potential for economic growth and re-industrialization. Hence, this is a period of dramatic economic change, and in many parts of the world, economic reform.

The global economic changes are accompanied by, and, in part driven by, a number of technological changes afoot. Bettis and Hitt (1995) trace four technological trends that are also shaping the competitive landscape facing organizations at the turn of the millennium. First is an increasing rate of technological change and diffusion. For example, the half-life of a computer system is now measured in weeks and months, no longer years or decades. Second is the way in which the current information age has produced information-rich, computation-rich and communication-rich work environments. For example, people at work have real-time access to vast amounts of global information at a moment's notice. Third is an increasing knowledge intensity resulting from technological change. For example, current knowledge intensity leads to a constant rate of change in a wide range of organizational expertises. Fourth is the emergence of the positive feedback industry. Positive feedback systems, for example, can be used to achieve continuous improvement in a wide range of organizational activities.

The human impact: threat-rigidity and stress effects

The economic and technological forces shaping the new competitive landscape have an important human impact, challenging individuals to adjust and accommodate in work as never before. While not all of the human impacts are negative, many are and there is always risk associated with change. Almost universally, individuals', groups' and organizations' first reflex response to change is to experience it as a threat, which in turn leads to varying degrees of rigidity on the part of the individual, group or organization (Staw, Sandelands and Dutton, 1981). This rigidity results from the restriction of information inputs from the environment and an increase in centralized control. Threat-rigidity effects may have functional or dysfunctional value for people at work. On the functional side, threat-rigidity effects enable people to respond and cope effectively with marginally changing work conditions. However, on the dysfunctional side, rigid responses to dramatically changed work environments may be inappropriate and/or inhibit the necessary learning required to respond appropriately to the changed environment.

The new organizational reality has second order effects in its human impact. These are the stress effects associated with the human risk that accompanies significant change. Cooper (1996) includes a set of four chapters which address the impact of life change events on risk of disease, with particular attention to cardiovascular disease and cancer. Life change events research has consistently shown over the years that change exposes people to the risk of a wide range of physical and psychological illnesses or diseases. Change places stress on the human system, demanding adjustment and accommodation. Change challenges the individual to maintain a condition of homeostasis with its associated experience of health and stability (Cannon, 1932, 1935). While some change is clearly valuable for enhancing growth and development, thus improving the overall capacity and competence of the person, too much change leads to excessive stress and increased risk of disease.

The human impact of excessive stress from the new organizational reality may become manifest in the increased risk of human illness, both physically and psychologically. However, there has been a sea change in the primary health risk exposures for people at work which has accompanied the dramatic industrial changes of the past several years. While industrial work and heavy manufacturing carry a wide range of physical risks for injury and accidents, service and information-based knowledge work carry an increased exposure to psychological risks, such as anxiety, depression and burnout.

Human distress falls into three broad categories: medical distress, psychological distress and behavioural distress (Cooper and Quick, 1999). Cooper's work has basically confirmed earlier physiological and endocrinological research on the risks of medical distress, for example cardiovascular disease and cancer, and extended this line of research into occupational contexts and with occupational stress exposures. What is increasingly significant in the new organizational reality is the human impact of psychological distress, most notably as manifest in depression, burnout, anxiety disorders, conversion reactions and work–family conflicts. Depression and anxiety are the two most common stress-related presenting complaints seen by general practitioners and family physicians. For example, anxiety disorders affect one in every six Americans and one in every five employed people in the United Kingdom (Cooper and Quick, 1999). Fortunately, the diagnosis and treatment of anxiety disorders has improved over the past decade.

Behavioural distress is the third form of human distress of concern to people in occupational work environments and associated with misguided attempts to cope with stress or traumatic stress events. Behavioural distress takes one or more of three dominant forms. These are substance abuse, violence and industrial accidents. The latter are the leading cause of death for men at work in the United States, while homicide is the leading cause of death for women at work (Quick, Quick, Nelson and Hurrell, 1997). However, these relatively low base rate events do not account for the non-fatal forms of human suffering resulting from behavioural distress. Substance abuse in the form of tobacco products may be the single most preventable cause of human suffering and death. One Federal Bureau of Investigation estimate suggests that over 90% of workplace violence is triggered by a stressful event, and is preventable (Mack, Shannon, Quick and Quick, 1998). Violence may be manifest either verbally or physically, and up to 25% of patients with stress-related problems may have suicidal thoughts which are not necessarily manifest.

A new age of self-reliance

Levinson (1981) brought attention to the problem of executive burnout nearly twenty years ago, as a relatively new phenomenon at the time. He characterized burnout as a special kind of exhaustion resulting from chronic stress and unremitted striving. Levinson argued in his classic article on burnout that prevention was the best cure. Thus, effective time management practices might help counteract overextension while planned changes in work schedules and/or assignments might provide healthy, stimulating stress.

In contrast to the early 1980s, Levinson (1996) argues that the 1990s was a decade in which feelings of stress are pervasive and growing worse. Corporate warfare, downsizing, re-engineering and restructuring were increasing the risks of combat stress, battle fatigue and other forms of distress in organizations (Nelson, Quick and Quick, 1989). The implicit psychological contracts of the 1980s between employees and their organizations in which the organization was seen as a source of security and psychological safety are now outdated. While employees used to be able to look to their organization to support them, these same employees must now look to themselves. Levinson (1996) goes on to suggest that one psychological and practical result of the new organizational reality is that we live now in a new age of self-reliance.

This has both personal and professional consequences. Personally, employees increasingly turn to personal resources, such as friends and family, as sources of support and psychological safety. Professionally, employees need to develop fallback positions as alternatives should the current work environment or organization fail. No longer can one develop a set of professional skills and rely upon them as an enduring source of self-reliance. Rather, one must think in terms of dynamic self-reliance as a continuous process of increasing professional mastery and enhancing competence (Frese, 1997).

The paradox of self-reliance

Levinson (1996) and Frese (1997) use the concept of self-reliance in the common colloquial manner understood by most people, which is as a term for describing a pattern of independent behaviour. This colloquial use of the term is at variance with the paradoxical pattern of interdependent behaviour originally described by attachment theories and subsequently by ourselves in a line of research over the past 16 years. According to early attachment theorists (especially Ainsworth and Bowlby, 1991), self-reliance has childhood origins. Subsequent research suggests that self-reliance as a paradoxical pattern of interdependent behaviour extends into the adulthood years. Attachment theory and its extensions are positioned on a developmental approach to human behaviour.

Childhood origins

Ainsworth and Bowlby were award-winning researchers who framed an ethological approach to personality based on attachment theory (Ainsworth and Bowlby, 1991). The core line of the theory describes attachment behaviour as one of two instinctual patterns of behaviour essential for healthy human development, the other instinctual pattern of behaviour being exploratory behaviour. As such, the instinctual predisposition to connect oneself to a secure attachment figure, such as a parent, in times of threat, danger and extreme stress has instrumental survival value for the child. However, healthy human development also dictates that the child engage in exploratory behaviour as well, going forth into the world to learn new behaviours and master new skills. While attachment behaviour and exploratory behaviour are somewhat antithetical in the moment, it is flexibility on the part of the child in shifting gears between these instinctual behaviours that leads to the fullness of growth and development which may be characterized as self-reliant.

Bowlby (1973) describes the problems and difficulties which may intrude on healthy human development when separation becomes an enduring characteristic of the childhood years. Separation from others accompanies exploration of the world and only becomes a developmental problem when one is unable to connect with the secure attachment figure in times of threat, danger and extreme stress. The inability to connect in time of need leads then to separation anxiety and subsequently anger. When separation anxiety and anger become a feature of the childhood years, there emerges a pattern of insecure attachment, in contrast to secure attachment. Insecure attachments can lead children to develop either dismissive or preoccupied patterns of attachment, which stand in contrast to a self-reliant and secure pattern of attachment.

A dismissive pattern of attachment is an insecure pattern rooted in a psychological and behavioural denial mechanism; that is, the child denies the need for other people in circumstances when that is appropriate. A preoccupied pattern of attachment is an insecure pattern rooted in the inability to connect and to experience felt security; that is, the child feels the need yet is unable to form an attachment accompanied by the experience of security and reassurance.

Bowlby (1988) originally described self-reliance as paradoxical because children who exhibited this pattern of behaviour acted with an apparent degree of independence and autonomy which belied their secure interpersonal attachments as bases of action. Thus, the self-reliant child appeared independent precisely because he or she was secure in their dependence upon others in time of threat, danger or extreme stress: the self-reliant child acted out of the experience of felt security.

Adulthood extensions

Ainsworth and Bowlby maintained a childhood focus in their attachment theory research over several decades, laying a solid foundation for other investigators (Ainsworth and Bowlby, 1991). Hazen and Shaver (1990) subsequently focused on the adulthood years, originally conceptualizing that attachment theory had implications for romantic relationships in the adulthood years. This first adulthood extension of attachment theory was very logical, based on the foundation research. About the same time they examined love and work as adulthood parallels to attachment behaviour and exploratory behaviour during childhood.

During the same era and in 1983, we commenced a line of successful executive research aimed at understanding how successful male and female executives who were healthy and vibrant maintain these characteristics in highly demanding and stressful circumstances. The basic interview protocol we used explored three aspects of the executive's world: his or her sources of stress, his or her early warning signs of distress and strain, and his or her stress management and coping strategies. The analyses of these information-rich interviews led us to an unanticipated finding. This was that the one common thread through all the interview data was the way in which these executives described their interpersonal supports at work and in their personal lives. While the stress management methods commonly described in the literature of the 1980s, such as exercise, prayer, meditation and time management, were used by some executives, the same methods were invariably not used by others. However, all had good interpersonal supports. The review process at the Academy of Management Executive connected us to attachment theory, and led us to clearly see how the pattern of behaviour revealed in the interviews was best interpreted through the lens of self-reliance (Quick, Nelson and Quick, 1987). We also saw the paradox of self-reliance in the overt autonomous and independent behaviour of the executives we studied, made possible by the network of secure attachments each had at work and at home.

Even in the adulthood years, self-reliance is a healthy, secure, interdependent behavioural strategy which may be characterized as a flexible, responsive strategy of forming and maintaining multiple, diverse relationships. The adulthood patterns of insecure attachment are counterdependence and overdependence, each of which is unhealthy. Counterdependence leads to the experience of separation and alienation while overdependence leads to the experience of intense separation anxiety and desperation. Counterdependence may be characterized as a rigid, dismissing strategy of denying the need for other people in difficult and stressful times. Overdependence may be characterized as a desperate, preoccupied strategy of attempting to achieve a sense of felt security through relationship.

Self-reliance research with executives

Cary Cooper became collegially and editorially supportive of the self-reliance research with executives. Specifically, he encouraged us to extend the 10 Self-Reliance questions from the original article into a 20-item scale with Likert-type responses and, concurrently, expand our treatment of the subject into a volume in his Wiley Series on Studies in Occupational Stress with Stan Kasl. The former led to the development of the Self-Reliance Inventory (Quick, Nelson and Quick, 1991), which was empirically refined into the SRI II under the leadership of Janice Joplin (Joplin, Nelson and Quick, 1999) and translated into German (SRI-G) by Joachim Strober and Spanish by Guillermina Garza Trevino. The latter led to *Stress and Challenge at the Top: The Paradox of the Successful Executive* (Quick, Nelson and Quick, 1990), elements of which were the core of our corporate warfare article (Nelson, Quick and Quick, 1989).

As we worked with Cary Cooper, we began to both understand and appreciate the way is which he embodies the pattern of self-reliance at the heart of the theory and research. One of the characteristics of self-reliance is the pattern of reciprocal interdependency which characterize the relationship involved. The core thesis of our self-reliance research is that such patterns of relationships ease the risks associated with the potential burdens of occupational stress.

Implications of self-reliance for social support, mentoring and health

Self-reliant individuals, those who exhibit the interdependent attachment style, should reap rewards from their ability to form secure, reciprocal relationships with others. The organizations they work in should benefit commensurately. There are three extensions of attachment behaviour that are of importance to both individuals and organizations. These extensions are social support networks, mentoring relationships, and health and well-being.

House, Landis and Umberson (1988) found that social isolation was a significant risk factor for both morbidity and mortality. Their meta-analysis demonstrated that positive health outcomes such as longer life span are associated with social support. Five additional prospective studies showed an increased risk of mortality associated with lower levels of social integration (Quick, Nelson, Matuszek, Whittington and Quick, 1996). The conclusion is a simple one: social isolation brings risks of ill health and death, while social support is healthy. Social support may be provided in many forms, including emotional support (empathy, concern), appraisal support (feedback and social comparison), informational support (advice, suggestions), and instrumental support (concrete aid such as money or time). A well-developed social support network consists of multiple forms of support and multiple providers.

The mechanism, however, whereby social relationships affect health is unclear. We believe that attachment style may be an important explanatory factor. In our studies of chief executives, it became clear that healthy executives are both providers and recipients of social support; they clearly exhibit the interdependent style of forming relationships with others. Self-reliant individuals are able to form secure relationships that are reciprocal in nature. They exchange resources, both material and interpersonal, with others as situationally appropriate. In essence, they transcend their own limitations by developing strong personal and professional support networks. A recent *Fortune* article described a study that shows the opposite side of this issue (Charan and Colvin, 1999). Researchers analysed notable failures of chief executive officers and the reasons CEOs fail. Their conclusion was that most CEOs possess the ability and the vision to succeed. Their failures stemmed from the inability to put the right people in the right jobs, and the failure to fix people problems in time. The CEOs failed because they did not develop an interpersonal infrastructure that complemented their own abilities.

The importance of well-developed social support networks goes beyond just the top level of the organization. Such networks are necessary for the health of all organization members. In our own study of organizational newcomers, we found that the availability of social support was associated with psychological well-being, and the helpfulness of social support was related to positive newcomer adjustment (Nelson and Quick, 1991). New organization members need support from a variety of sources if they are to succeed. In particular, support from the immediate supervisor and from co-workers is important. Interdependent newcomers are more likely to recognize the need for this support and to feel comfortable seeking it out.

Related to social support is the idea of mentoring. Cary Cooper has a reputation for mentoring young academicians and providing opportunities for them to learn and succeed. As editor of the *Journal of Organizational Behavior*, he took the unusual step (and perhaps the risk) of appointing young academicians to the editorial board, normally the bastion of more senior scholars.

Because self-reliant individuals form relationships that are reciprocal in nature, it stands to reason that such individuals would be attracted to the idea of mentoring other individuals. Mentors are important to career success because they provide both career (e.g. coaching, facilitating exposure) and psychosocial (e.g. role modelling, counselling) functions. Individuals with mentors have higher promotion rates and higher incomes than those who lack mentors, and they are also better decision makers (Scandura, 1992). The mentoring relationship is similarly beneficial to the mentor. Those individuals in the late phases of their own career find that mentoring re-energizes them and keeps them motivated. Mentoring relationships are by their very nature interdependent relationships. Fully developed mentoring relationships end up with the protégé and mentor becoming equals and peers. As the relationship matures, a natural separation process occurs. Self-reliant individuals, with an inner sense of felt security, can weather this transition in the mentoring relationship. Thus, we propose that self-reliant individuals can contribute significantly to the organization through mentoring relationships.

Perhaps the most important implication of self-reliance for both individuals and organizations relates to health and well-being. As a flexible, responsive strategy of building adult relationships, interdependence is a healthy characteristic. It can promote health and well-being on its own, or through the social support derived from the self-reliant person's well-developed network. Three studies are informative in relating attachment styles to health. Two studies were conducted in the military with basic trainees, one as a cross-sectional, comparative group analysis (n = 158) and one as a predictive validity study (n = 1339). Both studies showed that self-reliant trainees performed better in training than their overdependent and counterdependent peers. They were also healthier, had higher self-esteem, and lower burnout (Quick, Joplin, Nelson, Mangelsdorff and Fiedler, 1996). Another study involved 297 students, the majority of whom were also full-time employees. Relationships among the three attachment orientations, physical and psychological symptoms, and social support were examined. Interdependence was negatively related to social dysfunction.

Counterdependent and overdependent attachment styles were negatively related to health (Joplin, Nelson and Quick, 1999). We can conclude from this evidence that self-reliant individuals, who form interdependent relationships with others, enjoy better health than those individuals who are not self-reliant.

Given that self-reliant individuals have better-developed support networks, can serve readily as mentors, and have better health, there are things organizations can do to encourage self-reliance. Management development is particularly important because self-reliant managers can serve as referents and role models for others (Nelson, Quick, Eakin and Matuszek, 1995). The culture of the organization should be one in which employees are trained to diagnose work situations, rely on their own resources when appropriate, and ask for support from others when it is needed. Far too many organizational cultures encourage independence to the point that individuals are afraid that asking for help will be seen as a sign of weakness. Team building at all levels of the organization is another way to promote the development of interdependent relationships. Newcomers should be encouraged to develop social support systems, and special attention should be given to high risk groups such as people who travel frequently and those who tend toward counterdependence.

Four research issues

Although attachment theory has a long history in counselling and developmental psychology, it has only recently migrated into the organizational sciences literature. In addition, there is limited research concerning attachment styles in adult life. The possibilities for extending attachment theory into various areas of organizational studies are many. Four research issues that are especially intriguing are the developmental issues surrounding attachment, the work-related outcomes of attachment styles, the role of interpersonal trust in attachment relationships, and the idea of attachment to groups and organizations.

The developmental nature of attachment styles

One question surrounding attachment theory stems from its description as a developmental construct. Is attachment style stable throughout life, or is it malleable? Attachment styles are influenced by early experiences in the individual's family of origin, but how, and to what degree, are they influenced by current relationships? By relationships in the workplace? Some evidence does indicate different distributional patterns of attachment styles in groups of young and middle-aged adults compared with older adults (Diehl, Elnick, Bourbeau and Labouvie-Vief, 1998). A larger proportion of young and middle-aged adults described themselves as preoccupied or fearful (i.e. overdependent). Compared to young and middle-aged adults, a larger proportion of older adults described themselves as dismissing (i.e. counterdependent). The reasons behind these findings warrant exploration, and are most probably related to developmental issues.

Ultimately, self-reliant individuals develop a secure base within themselves. The process by which this occurs has not been explored in research. One way it might occur is through the use of dialogues as communication vehicles. Dialogues involve more than active listening to another person. They constitute a new way of paying attention to the self and others by observing one's own thoughts and feelings in a reflective way, and by engaging in collective reflection with others (Senge, Kleiner, Roberts, Ross and Smith, 1994). Dialogue is a way to become more mindful of the way we relate to others, to explore the reasons behind the styles of relating, and to potentially alter those styles. It holds promise as one way individuals can develop an internal secure base.

The outcomes of attachment styles at work

There are many unexplored outcomes of attachment styles at work. It is a natural and logical extension to predict that people who use an interdependent style in forming relationships are more productive, have more upward velocity in their careers, and will assume leadership roles as we move into the future. The notion of self-leadership, for example, blurs the distinction between leaders and followers. Self-leadership is embedded within the idea of self-reliance. The notion of what constitutes a career has evolved from one in which the organization takes care of the individual to one in which individuals take control of their own careers. In this new career paradigm, people who will succeed are those who are flexible, team-oriented, open to change, and tolerant of ambiguity. The internalization of secure working models of relationships with others should give self-reliant individuals an advantage here. To the extent that self-reliant individuals are more likely to possess these attributes, they can be expected to have more upward velocity in their careers.

Within the context of the new team environment, people are expected to come up with initiatives rather than follow orders (Hirshhorn, 1991). Team members both rock the boat and co-operate; they communicate through direct talk. This is in contrast to the old team environment, in which teams were directed by managers, and people suppressed thoughts and feelings.

We might also expect that self-reliant individuals are more apt to display organizational citizenship behaviours (OCBs) – behaviours that are above and beyond the call of duty in organizations. Such OCBs include helping co-workers, making positive comments about the company, and refraining from complaints when things do not go well at work. Future research should turn to the questions posed by the links between self-reliance and the above-mentioned constructs. The questions that evolve include the following: Are self-reliant leaders more effective? Do self-reliant individuals perform better at work? Are they more effective team members? Do they exhibit higher levels of organizational citizenship behaviours?

Attachment styles and trust

Trust is a 'hot' topic in organizational research, and it has strong roots in attachment theory. A recent review of cross-disciplinary scholarly writing offered the following as a widely held definition of trust: 'a psychological state comprising the intention to accept vulnerability based on positive expectations of the intentions or behaviour of another' (Rousseau, Sitkin, Burt and Camerer, 1998, p. 395). In infancy, the relationship with the primary caregiver provides trust in a secure base for self-reliant individuals. They are, therefore, more likely to possess positive expectations of another's behaviour.

There are several provocative research questions about the links between trust and attachment styles. Are self-reliant individuals perceived as more trustworthy by others? There are three factors that appear to be related to trustworthiness in organizations. These are ability, benevolence and integrity (Mayer, Davis and Schoorman, 1995). It would be interesting to explore these facets as they are related to self-reliance. Do self-reliant individuals have a higher propensity to trust others in the organization? When a psychological contract at work is breached, how do individuals with different attachment styles react? Are self-reliant individuals more forgiving of others they work with? All of these questions revolve around the relationship between attachment styles and trust.

Attachment styles have been categorized as belonging to the research stream of personality-based trust, in which trust develops during childhood. Two other research streams focus on institution-based trust and cognition-based trust (McKnight, Cummings and Chervany, 1998). Institution-based trust is based on the security the individual feels in a situation due to guarantees or social structures. Cognition-based trust is based on cues such as first impressions rather than interpersonal interactions. An interesting research question is how the various types of trust (personality-based, institution-based, cognition-based) develop over time and inform each other over an individual's life span.

Attachments to groups and organizations

Finally, research is needed that focuses on the role of interpersonal attachments to groups and organizations in the context of new work arrangements, such as troupes and virtual teams. The new organizational reality and the explosive use of information technology has made possible a wide range of working arrangements that do not depend upon face-to-face interaction or physical proximity. Mack (1996) suggests that the theatrical troupe may be a highly appropriate model for virtual work teams of the future. As in the case of a theatrical troupe, a work troupe might come together for a specific work project or assignment, then disperse until the individuals are then called upon again to be assembled into a new work troupe with a new or different task. In the case of virtual teams, individual members may never see, physically interact, or even telephonically communicate with each other during the course of a project. For example, a virtual team may complete their work interactively and electronically over the Internet. Mack (1996) has used the example of reserve officers at the Air Force Materiel Command's Aeronautical Systems Center who operate on many projects as troupes or virtual teams, in many cases performing virtual duty with work product deliverables without leaving their homes.

These new work arrangements and attachments call for research to examine their impacts on individuals' and team members' health and well-being as well as their impacts on group and organizational performance. For example, do troupes and virtual teams create difficulties even for employees with secure attachment styles? Do employees with insecure attachment styles have significantly more difficulty in completing their work in timely and predictable ways? How does attachment style (secure versus insecure) interact with the organizing forms (troupes and virtual teams) to effect security and predictability in the modern work environment? Finally, given the developmental nature of attachment styles, what are the effects of the new forms of work arrangements on the development and/or inhibition of secure forms of attachment?

References

Ainsworth MDS, Bowlby J (1991) An ethological approach to personality. American Psychologist 46: 333–41.
Bettis RA, Hitt MA (eds) (1995) Technological transformation and the new competitive landscape. Strategic Management Journal, Special Issue, 16: 7–200.
Bowlby J (1973) Separation. New York: Basic Books.
Bowlby J (1988) A Secure Base. New York: Basic Books.
Cannon WB (1932) The Wisdom of the Body. New York: WW Norton.

Cannon WB (1935) Stresses and strains in homeostasis. American Journal of the Medical Sciences 189: 1–14.

Charan R, Colvin G (1999). Why CEOs fail. Fortune June 21: 69–78.

Cooper CL (1996) Handbook of Stress, Medicine, and Health. Boca Raton, FL: CRC Press.

Cooper CL, Quick JC (1999) FAST FACTS: Stress and Strain. Oxford: Health Press.

Diehl M, Elnick AB, Bourbeau KS, Labouvie-Vief G. (1998) Adult attachment styles: Their relations to family context and personality. Journal of Personality and Social Psychology 74: 1656–69.

Frese M (1997) Dynamic self-reliance: An important concept for work in the twenty-first century. In: Cooper CL, Jackson SE (eds) Creating Tomorrow's Organizations: A Handbook for Future Research in Organizational Behavior, pp. 399–416. Chichester: Wiley.

Gowing MK, Kraft JD, Quick JC (1998) The New Organizational Reality: Downsizing, Restructuring, and Revitalization. Washington, DC: American Psychological Association.

Hazan C, Shaver P (1990) Love and work: An attachment-theoretical perspective. Journal of Personality and Social Psychology 59: 270–80.

Hirshhorn L (1991) Managing in the New Team Environment. Reading, MA: Addison Wesley.

House JS, Landis KR, Umberson D (1988) Social relationships and health. Science, 241: 540–5.

Joplin JRW, Nelson DL, Quick JC (1999) Attachment behavior and health: relationships at work and home. Journal of Organizational Behavior 20(6):783–96.

Levinson H (1981) When executives burn out. Harvard Business Review 59: 73–81. [Reprinted as an HBR Classic, Harvard Business Review 74: 153–63.]

Levinson H (1996) A new age of self-reliance. Harvard Business Review 74 162–3.

Mack DA (1996) Organization control for the nineties: Markets, bureaucracies, clans, and troupes. Proceedings of the Eleventh Annual Texas Conference on Organizations, pp. 83–92. Austin, TX: College of Business Administration, University of Texas at Austin.

Mack DA, Shannon C, Quick J D, Quick JC (1998) Stress and the preventive management of workplace violence. In: Griffin RW, O'Leary-Kelly A, Collins J (eds) Dysfunctional Behavior in Organizations: Violent and Deviant Behavior, pp. 119–41. Greenwich, CT: JAI Press.

McKnight DH, Cummings LL, Chervany NL (1998). Initial trust formation in organizational relationships. Academy of Management Review 23: 473–90.

Martin RE, Freeman SJ (1998) The economic context of the new organizational reality. In: Gowing MK, Kraft JD, Quick JC (eds) The New Organizational Reality: Downsizing, Restructuring, and Revitalization, pp. 5–20. Washington, DC: American Psychological Association.

Mayer RC, Davis JH, Schoorman, FD (1995) An integrative model of organizational trust. Academy of Management Review 20: 709–34.

Nelson DL, Quick JC (1991) Social support and newcomer adjustment in organizations: Attachment theory at work? Journal of Organizational Behavior 12: 543–54.

Nelson DL, Quick JC, Eakin ME, Matuszek PAC (1995) Beyond organizational entry and newcomer stress: Building a self-reliant workforce. International Journal of Stress Management 2: 1–14.

Nelson DL, Quick JC, Quick JD (1989) Corporate warfare: Preventing combat stress and battle fatigue. Organizational Dynamics 18: 65–79.

Quick JC, Joplin JR, Nelson DL, Mangelsdorff AD, Fiedler E (1996). Self-reliance and military service training outcomes. Journal of Military Psychology 8: 279–93.

Quick JD, Nelson DL, Matuszek PAC, Whittington JL, Quick JC (1995) Social support, secure attachments and health. In: Cooper CL (ed.) The Handbook of Stress, Medicine, and Health, pp. 269–287. Boca Raton, FL: CRC Press.

Quick JC, Nelson DL, Quick JD (1987) Successful executives: How independent? Academy of Management Executive 1: 139–45.

Quick JC, Nelson DL, Quick JD (1990) Stress and Challenge at the Top: The Paradox of the Successful Executive. Chichester: John Wiley.

Quick JC, Nelson DL, Quick JD (1991) The self-reliance inventory. In: Pfeiffer JW (ed.) The 1991 Annual: Developing Human Resources, pp. 149–61. San Diego, CA: University Associates.

Quick JC, Quick JD, Nelson DL, Hurrell JJ, Jr (1997) Preventive Stress Management in Organizations. Washington, DC: American Psychological Association. (Original work published in 1984 by JC Quick and JD Quick.)

Rousseau DM, Sitkin SB, Burt RS, Camerer C (1998) Not so different after all: A cross-discipline view of trust. Academy of Management Review 23: 393–404.

Scandura T (1992) Mentorship and career mobility: An empirical investigation. Journal of Organizational Behavior 13: 169–74.

Senge PM, Kleiner A, Roberts C, Ross RB, Smith BJ (1994) The Fifth Discipline Fieldbook. New York: Doubleday.

Staw BM, Sandelands LE, Dutton JE (1981) Threat-rigidity effects in organizational behavior: A multilevel analysis. Administrative Science Quarterly 26: 501–24.

Chapter 3
Stress and Health – from a Work Perspective

Töres Theorell

Introduction

This chapter has a medical approach and therefore the physiological defini-tion of stress that was introduced by Selye will be used. Hans Selye intro-duced the stress concept in medicine in the 1950s. According to his way of defining it, stress is a general response pattern that is triggered by many different kinds of stressors, psychological as well as physical. Although different individuals react differently to one and the same situation and one individual reacts differently to different situations – stressors – there is a common denominator of all these reactions and this is what Selye labelled stress. This common denominator is energy mobilization, which is needed in many situations that human beings are facing. Phylogenetically it was very important to be able to suddenly and rather infrequently mobilize substan-tial energy in physically dangerous or demanding situations. In the present world of work, however, the patterns of energy mobilization are quite different. Physical demands have become infrequent. Socially and psychoso-cially demanding situations on the other hand are becoming increasingly common. Four hypotheses will be discussed in this review.

1. *Increased exposure to socially demanding situations*. In the modern work situation there are an increasing number of social contacts and situations in which social functioning is needed. This is partly due to the fact that work is becoming more and more specialized (which means increased dependency on other specialists) and partly due to the increasing use of information technology (which increases the number of contacts between people). Many of these contacts will be psychologically demanding since many of them are novel to the person. Jobs with few social contacts are becoming increasingly rare.

2. *Increased frequency of disorders of regulation of energy mobilization.* Because of the increasing number of socially demanding situations energy mobilization takes place increasingly often, and this means that disorders of regulation of energy mobilization are becoming more frequent. This may result in increasing rates of work-related depression and various forms of serious exhaustion.

3. *From paralysis of muscles to paralysis of memory.* Since cognitive demands on all workers are increasing, cognitive functions are replacing the role that physical functions have had historically in the world of working. Conversion hysteria has been common as a psycho-somatic disorder in the past – this corresponds to a non-organic (non observable anatomic changes in the structures) paralysis of an arm or a leg of particular significance to their working. Today paralysis of memory or inability to perform cognitively may have replaced muscular paralysis. In various forms of exhaustion cognitive disturbance is a common feature, for instance in chronic fatigue syndrome.

4. *Increasing importance of severe psychological trauma.* Workers who have experienced extraordinarily traumatic events are becoming more prevalent in the modern working world. They have special difficulties related to energy mobilization since they are physiologically sensitized to the memories of this traumatic experience. Whenever they are reminded of this their body mobilizes substantial amounts of energy that are not needed. This creates difficulties in concentrating and functioning in the modern workplace.

Acute and long-term stress reactions

In the first phases of stress research almost all scientific effort was devoted to the description of acute stress reactions in situations that demanded energy mobilization. During later years, however, the emphasis has been on the effects of long-term stress on the regulation of arousal mechanisms. Researchers have been puzzled by the fact that psychosocially adverse conditions at work are frequently not associated with high plasma concentrations of cortisol despite the fact that such an association would be expected. A constantly elevated plasma cortisol concentration, however, may influence the regulation of arousal in several ways.

Break failure

In a healthy situation an elevated plasma cortisol concentration should inhibit the upper 'driving' parts of the HPA (hypothalamo-pituitary-adrenocortical) arousal system. If the system is overused, this inhibitory mechanism may be functioning less well. The result could be described as 'break failure'. That such a regulatory disorder exists has been known for a long time (see Rubin, Poland, Lesser, Winston and Blodgett, 1987). It corresponds to psychiatric depression in which plasma cortisol is constantly elevated. Inhibition by means of dexamethasone (a synthetic analogue of cortisol) does not work. It is not known to what extent overuse of the arousal system may induce break failure. The extent to which long-term arousal contributes to this condition is unknown.

Accelerator failure

The HPA axis may also be passive and non-reactive to external stressors. This means that the normal activation does not take place when the person is confronted with challenging or threatening situations, for instance at work. This could mean being non-reactive both physically and emotionally. Such a pattern has been shown to be frequent in patients who have the diagnosis chronic fatigue syndrome (CFS) (Demitrack, Dale, Strauss, Laue, Listwak, Kruesi, 1991). Recently it has also been shown to be common in the metabolic syndrome (with increased blood pressure or abdominal adiposity or type 2 diabetes or combinations of these conditions) (Rosmond, 1998). In these disorders the normal circadian cortisol variation – with high morning concentrations and successively lower ones during the waking hours – has been flattened and the levels are in general lower than normal. In the following argument this condition will be labelled 'exhausted HPA axis'.

The extent to which long-term arousal contributes to exhausted HPA axis is unknown. It is not unlikely that physical stressors may contribute as well. CFS has been assumed to be induced by either repeated infections or chronic psychosocial arousal or a combination of these (Wessley, Chalder, Hirsch, Wallace and Wright, 1997). It has been hypothesized that other parallel biological systems such as the sympatho-adreno-medullary system (producing noradrenaline and adrenaline) may be hyperreactive when the HPA axis is hypoactive – in order to compensate for the loss of arousal ability. In a similar vein, hyperactivity in immune functions has been found in CFS (Kavelaars, Knook, Prakken, Kuis and Heijnen, 1997). It is not known, however, what the direction of causality is. Are repeated infections making the immune system hyperactive and concomitantly exhausting the HPA axis? Or is it the other way around, that an exhausted HPA axis (perhaps due to long-term psychosocial arousal) has to be compensated for by overactivity in the immune system? Or are they unrelated parallel processes?

Mixtures of and/or transitions between 'break failure' and 'accelerator failure' have not been studied – and are of course very difficult to study in humans. There is a possibility that they may be stages in a development. It is possible, for instance, that psychiatric depression with 'break failure' precedes 'exhausted HPA axis'.

One of the most common complaints in chronic fatigue syndrome is impaired cognition (memory above all) and this, of course has a profound impact on working capacity. It is possible that most of the cognitive difficulties are due to the fatigue itself and that no objective loss in cognitive ability has taken place. This has to be subjected to research because it is very important for workers in the working world of today and in the future. The demands on cognitive functions are constantly increasing because of the increased use of information technology and computers.

Hypersensitivity in the HPA axis

The experience of extraordinarily severe trauma is associated with the development of post-traumatic stress disorder (PTSD) (Michelsen, Licinio and Gold, 1995). In this condition the sensitivity of the cortisol receptors as well as the number of receptors is elevated. This means that whenever the subject is reminded of the traumatic event he or she develops an extremely aroused condition. Anything that is a reminder of the event may arouse this state, and this may happen many times a day. This is associated with markedly impaired ability to concentrate, which is incapacitating at work, and there are an increasing number of people with PTSD at work in the industrialized world. Neurobiological studies have shown that there may even be structural changes in one part of the brain, the hippocampus, which has an important role in the memory function. These changes may be reversible. Their existence indicates that severe stress may induce biological brain changes that may partly explain the cognitive difficulties of subjects with PTSD (Bremner, Randall, Scott, Bronen, Seibyl, Southwick, Delaney, McCarthy, Charney and Innis, 1995; Gurvits, Shenton, Kokama, Ohta, Lasko, Gilbertson, Orr, Kikinis, Jolesz, McCarley and Pitman, 1996). The significance of this to working capacity should be subjected to research.

Since bullying at work is a relatively common phenomenon and the traumatic events associated with it may induce PTSD-like conditions (Leymann, 1990), PTSD may be of greater importance to the relationship between health and work than has been previously realized. So far PTSD has been relatively unrecognized in relation to work. Another reason why PTSD may be relevant to work environment researchers is that 'indirect' trauma may also induce PTSD-like conditions. In health care work, witnessing dramatic unexpected deaths or being guilty of errors that cause severe complications for patients may induce PTSD-like conditions. According to an Australian study the prevalence of these conditions may be in the order of 18% of health care workers (King, 1999).

The old diagnoses

There is a new discussion regarding the relative importance of the 'old' risk factors for coronary heart disease. It is possible that the three most important classical risk factors – smoking, elevated blood pressure and lipoprotein abnormalities – may not explain as much of the risk factor pattern as had previously been thought. Risk factors that seem to have gained importance during recent years are physical activity during leisure as well as 'stress at work'. For both of these risk factors it is possible that improved measurements in epidemiological studies could explain part of the change. It is also possible, however, that the observed changes may reflect true changes in the importance of these factors. Body weight in relation to height (body mass index, BMI) is increasing in many of the industrialized countries ((Wilhelmsen, Johansson, Rosengren, Wallin, Dotevall and Lappas, 1997). This may be a consequence of decreased physical activity both at work and during leisure – which may increase the relative importance of this risk factor. Tobacco consumption has decreased in most industrialized countries, and this may have decreased the relative importance of tobacco in the total risk pattern. With regard to stress at work the development has both positive and negative components, as described below.

Several theoretical models have been used in the study of stress at work and cardiovascular disease. The first model used was the person–environment fit model (Katz and Kahn, 1966). Cooper's well-known occupational stress inventory has also been used in large studies of cardiovascular risk factors and symptoms of heart disease (Cooper and Marshall, 1976). During recent years Siegrist's effort–reward imbalance model has also been used extensively in relation to coronary heart disease (Siegrist, 1996).

Empirical tests of the demand–control model

Karasek's original hypothesis, that the combination of excessive psychological demands and lack of decision latitude (control) is associated with increased risk of cardiovascular disease, has been tested in a large number of epidemiological studies in the 1980s and 1990s. There have been many published prospective or cross-sectional studies. The methodology has varied considerably. The most important distinction between the methodologies is that some studies have used the victims' own descriptions of their work situations, whereas others have used aggregated job descriptive data, based on representative workers in the occupations in the population. Both methods have advantages and problems. For example, individual traits may be associated with systematically distorted work descriptions, and this systematic distortion may be related to illness risk – with both overestimation and underestimation of the relative risk as a possible result. The use of aggregated data avoids individual distortion (although, of course, collective distortion may still take place). On the other hand, the use of aggregated data does not allow for variations between work sites. This may lead to substantial underestimation of true associations (Alfredsson 1983). The underestimation problem in the use of aggregated data is probably pronounced in estimating the importance of psychological demands, since this variable shows relatively small variance between occupations. Decision latitude shows considerable variance between occupations (Karasek and Theorell, 1990). Even the aggregated methodology has varied across studies. In some studies, the classifications have been based on means for each one of the dimensions from employees in the different occupations in the working population. In others, single questions have represented the dimensions, and several combinations (demand/skill utilization and skill/authority over decisions) have been tested. Finally, in a third group of studies, expert ratings have been used as a way of assessing working conditions objectively.

Studies that have used the two dimensions together have for the most part provided better predictions than studies using either one of them alone. At the same time, decision latitude has been of greater significance empirically in most studies than have psychological demands. In several studies during recent years it has proved difficult to operationalize and conceptualize psychological demands for epidemiological studies. This may be the main reason why studies have shown inconsistent results with regard to the association between psychological demands and heart disease risk.

The summary of relative risks indicates, as expected, that studies utilizing self-reported work descriptions have shown higher and more statistically consistent relative risks than have studies utilizing aggregated measures. Some of the studies have covered other risk factors including personality factors. In general the adjustment for standard risk factors for cardiovascular disease does not eliminate the association between the high demand–low decision latitude combination and cardiovascular disease risk. In fact, in one case, the Framingham study (La Croix 1984), the adjustment for other risk factors strengthened the association.

Studies of participants younger than 55 years of age in general have typically shown stronger associations than those including older subjects (Theorell and Karasek, 1996). Psychosocial job conditions may be of less importance during the ten later years of the working career than before that period. The patterns are probably different for men and women. Another observation is that the high demand–low decision latitude combination has proven to be a more powerful predictor of cardiovascular illness risk in blue-collar than in white-collar men. For instance, a Finnish study (Haan, 1985) included mainly blue-collar workers and showed a strong association. The study by Johnson and Hall (1988) includes separate analyses of blue-collar and white-collar men which illustrate this point. The SHEEP (Stockholm Heart Epidemiology Program) study has shown that the job strain factor has better predictive value for blue-collar than for white-collar workers (Theorell, Tsutsumi, Hallqvist, Reuterwall, Hogstedt, Fredlund, Emlund and Johnson, 1998). For further discussion see Marmot and Theorell (1988). Hallqvist has made the point (Hallqvist, Diderichsen, Theorell, Reuterwall and Ahlbom, 1998) that there may be synergy between social class and job strain. According to the empirical findings in that study the effect of job strain may be more pronounced in blue-collar workers than in white-collar workers.

There have also been several studies with lack of findings for both demand and decision latitude. The following characteristics seem to be common in such negative studies:

1. Long follow-up periods. A single measure of job conditions at one point of time is not likely to be predictive for periods longer than five years. Workers may change jobs and conditions may change too.
2. Indirect 'aggregated' measures of job conditions. In particular, aggregated measures based upon working populations in the 1960s and 1970s may not be relevant. Such aggregated measures have to be reconstructed as the labour market changes.
3. Older study populations, in particular when participants are older than 55 years from the start. In this type of study, many of the participants retire during follow-up.
4. Study populations with little variation in decision latitude, such as groups that have only one type of white-collar workers. In such a study the only 'available variance' is in *perception* of working conditions (which the study is not designed to test).
5. Case samples with coronary heart disease of varying duration preceding the examination. This factor may lead to adaptation to the illness which may cause those subjects who have the most pronounced symptoms of illness to select easier jobs and improved conditions. This could cause serious underestimation of the causal role of job conditions.

Another important methodological issue has to do with the question whether self-reports correspond to more objective measures. A related question is the following: Do subjects who have recently suffered an acute illness describe their work situation in a way that is systematically different from subjects in the normal working population? In the SHEEP study in Stockholm, 'objective' (obtained by means of the Swedish job exposure matrix, JEM) and 'subjective' (self-reports) job descriptions were compared (Theorell et al., 1998). For decision latitude, the cases and the referents were very similar with regard to job reporting behaviour (the way in which self-reports and JEM assessments related to one another), although as expected there was only a moderate correlation between the two sources of information. The conclusion was that (in this case male) subjects who had had a recent myocardial infarction (interviewed within one month after the event) did not show 'differential recall bias' (memory distortion that is systematically different in the two groups) in relation to decision latitude. This means that systematic 'after-construction' on the part of the patient ('my infarction was caused by too little say at work') is an unlikely explanation of the relationship between low decision latitude and elevated risk of myocardial infarction in case-referent studies of this kind. The same was found for psychological demands, although in this instance the findings were weaker since the aggregated measure for demands was methodologically poor. Controlling for chest pain preceding the first myocardial infarction did not change the conclusions in this study either – which is further support for the conclusion that differential recall bias is not a significant problem. However, if subjects had had a long-lasting serious debilitating coronary heart disease preceding the examination the situation might be markedly different with resulting distortion of self-reports.

One of the important theoretical questions in this field is whether the psychological demands variable interacts with decision latitude in generating increased risk. A few attempts at elucidating this question have been made (e.g. Alfredsson and Theorell, 1983; Johnson and Hall, 1988; Reed, la Croix, Karasek, Millera and McLean, 1989). Most studies have not addressed this question specifically. Recently, Hallqvist et al. (1998), on the basis of the SHEEP study, showed that there may be a strong interaction between psychological demands and decision latitude in relation to myocardial infarction risk.

Thus, although many studies have not confirmed the expected interaction between psychological demands and decision latitude in relation to the main outcome variable, myocardial infarction risk, recent research indicates that this is a subject that should be explored more in detail in future research.

In coming years the question regarding how to operationalize (i.e. define within the context of measures that are used in the studies) and measure psychological demands at work must receive increased attention. There is a need to establish new concepts that are relevant for the assessment of demands in a changing world of work.

So far, fewer studies have been performed on women than on men. There is no indication, however, that job strain is less important to women than to men (Reuterwall, Hallqvist, Ahlbom, de Faire, Diderichsen, Hogstedt, Pershagen, Theorell, Wiman and Wolk, 1999). Hall (1990), in a study of random samples of Swedish working men and women, showed that the interaction between activities outside work (unpaid work) and work activities was more important for women than for men in relation to psychosomatic symptoms. As a consequence, the pattern of associations between psychosocial factors and health was different in men and women. Hall furthermore pointed out that men and women, in general, work in different kinds of occupations. This may also be a reason why the patterns of association are gender specific.

Social support added to the demand-control model

Good social support at work could be a protective factor for myocardial infarction. A study of cardiovascular disease prevalence in a large random sample of Swedish men and women indicated that the joint action of high demands and lack of control (decision latitude) was of particular importance to blue-collar men whereas the combined action of lack of control and lack of support is more important for women and white-collar men (Johnson and Hall, 1988). The relative importance of these three components may accordingly be different in different strata of the population. The effect of the interaction between all the three of them (iso-strain) was tested in a nine-year prospective study of 7000 randomly selected Swedish working men. For the most favoured 20% of men (low demands, good support, good decision latitude) the progression of cardiovascular mortality with increasing age was slow and equally so in the white-collar and blue-collar workers. However, the age progression was much steeper in the group of least favoured blue-collar workers, than it was in the group of least favoured white-collar workers (Johnson, Hall and Theorell, 1989).

Working life career

Attempts are being made to use the occupational classification systems in order to describe the 'psychosocial work career'. Researchers have pointed out (House, Strecher, Metzner and Robbins, 1986) that an estimate of work conditions only at one point in time may provide a very imprecise estimation of the total exposure to adverse conditions. Three digit job titles are obtained for each year during the whole work career for the participants. Occupational scores derived from national samples of other subjects are subsequently used for a calculation of an indirect measure of 'total lifetime exposure' – for the study of cumulative effect. The 'total job control exposure' in relation to nine-year age-adjusted cardiovascular mortality in working Swedes was studied in this way. It was observed for both men and women that the cardiovascular mortality differences between the lowest and highest quartiles were almost two-fold even after adjustment for age, smoking habits and physical exercise (Johnson, Stewart, Hall, Fredlund and Theorell, 1996).

Not only cumulative effects but also variation in job control may be significant. A recent study (Theorell et al., 1998) has shown that the level of control inferred from the job title – after taking age, gender and time of exposure to the occupation into account – has a different development in working men who have experienced a first myocardial infarction compared to a referent group of age-matched men without this experience. Thus, in the group of men developing the first myocardial infarction, decreasing levels of control were seen, particularly during the five years preceding the illness. This observation may illustrate that the timing of a first myocardial infarction in a working man may be related to falls in control level at work. In the near future there will be increasing numbers of lay-offs and changes in jobs – individuals will have to accept jobs with much lower levels of decision latitude than they have been used to. An important observation from the research, however, was that decrease in control was much more important for men below age 55 than it was for older men. For women decreased decision latitude was not a significant risk factor. The finding in men is in line with the observation in several studies that the *level* of decision latitude is also more important for men below age 55 than for older men.

The decrease in the importance of decision latitude to the risk of developing myocardial infarction after age 55 may be explained by the progression of decision latitude with age. According to a study of nearly 6000 working men and women in Stockholm, decision latitude increases rapidly during the first years of the career. In middle age (around 55 in men and around 45 in women) the peak is reached and during later years of the career, a slight decrease is observed (Theorell, 2000). Accordingly, in future studies it will be important also to relate stages in the life career to the effect of decision latitude and of changes in decision latitude.

Loss of control may be inevitable but the knowledge that negative change in decision latitude is associated with increased risk of developing a myocardial infarction emphasizes the need for psychological and social care-taking in situations of this kind. Our data indicate that the elevation of risk does not arise immediately – there is time for an engaged company, friend or care-giver to strengthen self-esteem.

Effort–reward imbalance

Siegrist's group has analysed the association between effort–reward imbalance and coronary heart disease risk in several epidemiological studies and Siegrist (1996) has summarized the findings. A recently published study of men and women in the Whitehall study of government employees in London showed that decision latitude (lack of control) and lack of effort–reward imbalance both contributed independently of one another to the prediction of new episodes of coronary heart disease in men and women, even after adjustment for a number of biological risk factors and social class (Bosma, Peter, Siegrist and Marmot, 1998). In this study psychological demands had no significant effect on risk. Effort–reward imbalance thus seems to be a factor that should be studied along with demand–control–support. The Whitehall analysis could be criticized for having included 'intrinsic effort', the term used for the individual's own drive in setting the demand level in the measurement of 'effort'. This inclusion makes the measure incommensurable with the measures used in the demand–control model. In recent analyses of Swedish epidemiological samples 'intrinsic effort' has been analysed as a separate variable. Another criticism of the Whitehall analysis has been that the sample of state employees studied was not representative of the total working force, one of the characteristics of the Whitehall samples for instance being that there is a relatively high positive correlation between psychological demands and decision latitude. This may make it difficult to study the effect of psychological demands on cardiovascular risk in that sample. An unpublished analysis of a large more representative case-referent study in Stockholm (SHEEP) showed that both job strain and effort–reward imbalance contributed independently to the prediction of cardiovascular risk in men in Stockholm.

Shift work, a special psychological demand

Adverse working hours could be psychologically demanding. Accordingly it is important to study the relationship between working hours and coronary heart disease risk. Constant rotation between night and day work, mostly labelled shift work, is associated with increased risk of developing a myocardial infarction (Knutsson, 1989; Knutsson, Hallqvist, Reuterwall, Theorell and Åkerstedt, 1999). Relative risks of the same order as those found for job strain have been found, particularly after many years of exposure. In one cohort study decreasing risk was observed in workers who had been shift workers for more than 20 years. Both pathophysiological mechanisms and associations between shift work and lifestyle (Knutsson et al., 1999) have been discussed as links between shift work and myocardial infarction risk.

Intermediating mechanisms

The question is how the relationship between 'iso-strain' and risk of cardiovascular illness arises. Part of it could be due to the fact that there is a relationship between accepted lifestyle risk factors such as smoking habits and job strain (Karasek and Theorell, 1990; Green and Johnson, 1990). Part of it could also be direct effects of an adverse job environment on biochemical and endocrinological factors.

Interesting results have been found with regard to the association between job strain and blood pressure. Thus job strain is relevant to blood pressure levels primarily during working hours but not to blood pressure during other parts of the day and night. However, after long-lasting exposure to an adverse job environment, the effect may also spread to the whole day and night, including during sleep. Most researchers have found a positive relationship between job strain and ambulatory blood pressure measurements, particularly during working hours. In a study of women, a clear association was also observed between job strain and plasma prolactin levels when the subjects arrived at work in the morning; plasma prolactin was also significantly associated with blood pressure levels at work. Schnall, Pieper, Schwartz, Karasek, Schlussel, Devereux, Ganan, Alderman, Warren and Pickering (1990) found very clear relationships between job strain and hypertension, and for the younger participants in their study between job strain and ventricular hypertrophy (increased thickness of the myocardium), even after adjustment for a number of potential confounders.

Blood lipid levels have not been studied extensively in relation to the demand–control–support model. One study has shown a relationship between job strain and low levels of high density lipoprotein in blood in farmers, however. A recent study of nearly 6000 working men and women in Stockholm (Alfredsson, Hammer, de Faire, Hallqvist, Theorell and Westerholm, 1997) showed a relationship between a high ratio of low density lipoprotein (LDL) to high density lipoprotein (HDL) cholesterol (which is known to be an 'atherogenic' index) on the one hand and job strain on the other, particularly in young men. Recently a clear relationship was observed between low decision latitude and suppression of HDL cholesterol in a random sample of Swedish working women (Wamala, Wolk, Schenk-Gustavsson and Orth-Gomér, 1998). A study by Siegrist, Matschinger, Cremer and Seidel (1988) of industrial workers has shown a relationship between chronic job stressors (such as threat of unemployment, lack of promotion possibility and shift work) and a high LDL/HDL ratio. A recent study has shown similar findings with regard to hypertension and atherogenic lipids in a large sample of Swedish workers (Peter, Alfredsson, Hammar, Siegrist, Theorell and Westerholm, 1998), with different patterns for men and women.

Thus, there may be a relationship between some aspects of job stress and suppression of the blood concentration of high density lipoprotein which is protective against atherosclerosis.

Blood coagulation has also been discussed in relation to psychosocial job conditions. Relationships were found between low control and high fibrinogen in the Whitehall study, and this has been verified in later studies of the same cohort (Brunner, Davey Smith, Marmot, 1996). Similar findings were made in a recent Swedish study (Tsutsumi, Theorell, Hallqvist, Reuterwall and de Faire, 1999).

The reason why I have discussed recent findings in the research on psychosocial job factors in relation to coronary heart disease is that this research has relevance for predictions of the development of coronary heart disease risk in the future.

If we accept the notion that the psychosocial job factors that have the most consistent relationship to coronary heart disease risk are (1) low decision latitude (with both little authority over decisions and little skill discretion), (2) high job strain (high psychological demands and low decision latitude), (3) poor effort–reward balance, with little reward for high effort and (4) shift work with periods of night work mixed with periods of day work, we may make predictions regarding the future role of psychosocial job factors in the incidence of coronary heart disease. In the part of the world that has been industrialized for a long time (northern Europe, North America, Japan, Australia and New Zealand) we expect a continued decrease in tobacco smoking, increased treatment of hypertension (resulting in normalized blood pressure levels) and increasing prevalence of overweight associated with decreased physical activity. During a long period from the early 1970s to the mid-1990s there has been a decreasing incidence of coronary heart disease, which may be explained first of all by the decrease in tobacco smoking in these parts of the world, since this risk factor has the strongest explanatory value. In the developing part of the world, on the other hand, tobacco smoking may continue to increase. Treatment of hypertension will increase but physical activity will decrease. Thus changes in biological risk factors will be more favourable in the 'old' industrialized world.

There is reason to believe that pronounced improvement in two of the major biological risk factors, smoking and blood pressure treatment, may have masked possible effects of declining psychosocial conditions. In Sweden there is clear evidence from population surveys (Statistics Sweden, 1997) that the prevalence of fatigue, severe anxiety and sleep disturbance has increased during the 1990s. This is associated with increased working hours and psychological demands. Employment security is much lower since unemployment is much more frequent now than during the preceding decades and temporary employment has become much more common (Aronsson and Sjögren, 1994). With regard to decision latitude, the 1970s and 1980s were associated with a pronounced improvement. During the 1990s, a split development has taken place with a *favoured* group with high education and stimulating jobs with good possibilities to influence the job situation and a *disfavoured* group with low educational levels, temporary employment in jobs with few learning opportunities and little possibility to influence decisions. The gap between these groups is widening both with regard to psychosocial work environment and coronary heart disease incidence (Hallqvist, 1998). At the same time, psychological demands and working hours are rising in the whole industrialized world (e.g. the USA, Kasper, 1999). This is a general phenomenon which affects both the favoured and the disfavoured groups.

Our results indicate that there may be a delay – of perhaps five years – between loss of job control and increased risk of developing myocardial infarction. This could mean that decreased decision latitude in the large disfavoured groups could result in increasing myocardial infarction incidence in the near future in these groups.

Social support at work, which is a less extensively examined risk factor than job decision latitude, is expected to deteriorate because of the increasing numbers of temporary jobs, which make it difficult to develop steady social relationships at work.

So far, psychological demands at work have not been as clear a risk factor for coronary heart disease as decision latitude. In fact the findings regarding psychological demands as a single risk factor have been inconsistent. There may be several reasons for this, for instance that the assessments of psychological demands have been poor. But an important reason could also be that there may be a threshold effect that has not been reached previously. It is only when we reach very high demand levels that they really make a difference, particularly when decision latitude, support and reward are poor. Perhaps we are starting to push workers beyond this limit. The 'new' increasing prevalence of fatigue and new diagnoses such as chronic fatigue may be indications that we are pushing people too hard with too high demand levels. If the 'new' physiology is right modern life is inducing physiological disturbances with an exhausted 'normal activation mechanism' (HPA axis). Counterparts of this have certainly existed in the past (Ursin, 1997), but it seems to be increasing in the working population. The fact that the normal activation mechanisms do not function may activate the body to use alternative more pathological activation mechanisms and this could increase the risk for cardiovascular disease. Appels and his co-workers have pointed out that this mechanism may be important and according to them 'vital exhaustion' (which is related to chronic fatigue syndrome) is associated with a pronounced elevation of the risk of developing myocardial infarction (Appels, 1997).

Concluding remarks

There is empirical evidence from population surveys in the USA and Sweden (Statistics Sweden, 1997; Kasper, 1999) that excessive working hours are more common today than they were ten years ago. Furthermore, psychological demands are increasing according to Swedish population surveys. Undoubtedly the numbers of computers and the demands on memory functions are increasing. Therefore, with regard to mental health, it is not surprising that problems with cognition are becoming increasingly common among workers. Most of this is probably not caused by objective changes in memory functions but is merely a combination of the increasing prevalence of fatigue and increased demands. In PTSD-like conditions the situation may be different since there is evidence that there are objective changes in the hippocampus, which has a central significance to the memory.

For cardiovascular disease the clinical manifestations are essentially unchanged but the panorama of risk factors may be changing, and in the near future the importance of physiological fatigue in the aetiology of coronary heart disease may increase.

References

Alfredsson L (1983) Myocardial infarction and environment: use of registers in epidemiology. Academic thesis, Karolinska Institutet, Stockholm, Sweden.

Alfredsson L, Hammer N, de Faire U, Hallqvist J, Theorell T, Westerholm P (1997) Job strain and major risk factors for coronary heart disease: baseline results from the WOLF study. Unpublished ms, Institute for Environmental Medicine, Karolinska Institutet, Stockholm, Sweden.

Alfredsson L, Theorell T (1983) Job characteristics of occupations and myocardial infarction risk: Effects of possible confounding factors. Social Science and Medicine 17: 1497–503.

Appels A (1997) Exhausted subjects, exhausted systems. Acta Physiologica Scandinavica Suppl. 640: 153–4.

Aronsson, G, Sjögren A (1994) Samhällsomvandling och arbetsliv. Omvärldsanalys inför 2000-talet. Fakta från Arbetsmiljöinstitutet.

Bosma H, Peter R, Siegrist J, Marmot M (1998) Two alternative job stress models and the risk of coronary heart disease. American Journal of Public Health 88(1): 68–74.

Bremner JD, Randall P, Scott TM, Bronen RA, Seibyl JP, Southwick SM, Delaney RC, McCarthy G, Charney, DS, Innis RB (1995) MRI-based measurement of hippocampal volume in patients with combat-related posttraumatic stress disorder. American Journal of Psychiatry 152(7): 973–81.

Brunner E, Davey Smith G, Marmot M (1996) Childhood social circumstances and psychosocial and behavioural factors as determinants of plasma fibrinogen. Lancet 347: 1008–13.

Cooper CL, Marshall J (1976) Occupational sources of stress: A review of the literature relating to coronary heart disease and mental ill health. Journal of Occupational Psychology 49: 11–28.

Demitrack MA, Dale JK, Straus SE, Laue I, Listwak SJ, Kruesi MJP (1991) Evidence for impaired activation of the hypothalamic-pituitary-adrenal axis in patients with chronic fatigue syndrome. Journal of Clinical Endocrinology and Metabolism 73: 1224–34.

Green KL, Johnson JV (1990) The effects of psychosocial work organization on patterns of cigarette smoking among male chemical plant employees. American Journal of Public Health 80: 1368–71.

Gurvits TV, Shenton ME, Kokama H, Ohta H, Lasko NB, Gilbertson MW, Orr SP, Kikinis R, Jolesz FA, McCarley RW, Pitman RK (1996) Magnetic resonance imaging study of hippocampal volume in chronic, combat-related post-traumatic stress disorder. Biological Psychiatry 40: 1091–9.

Haan M (1985) Job strain and cardiovascular disease. A ten-year prospective study. American Journal of Epidemiology 122: 532–40.

Hall EM (1990) Women's work: An inquiry into the health effects of invisible and visible labor. Academic thesis. Karolinska Institutet, Stockholm, Sweden.

Hallqvist J, Diderichsen F, Theorell T, Reuterwall C, Ahlbom A (the SHEEP Study Group) (1998) Is the effect of job strain on myocardial infarction risk due to interaction between high psychological demands and low decision latitude? Results from Stockholm Heart Epidemiology Program (SHEEP). Social Science and Medicine 46(11): 1405–15.

Hallqvist J (1998) Socio-economic differences in myocardial infarction risk. Epidemiological analysis of causes and mechanisms. Academic thesis. Karolinska Institutet, Stockholm, Sweden.

House JS, Strecher W, Metzner HL, Robbins C (1986). Occupational stress and health among men and women in the Tecumseh Community Health Study. Journal of Health and Social Behavior 27: 62–77.

Johnson JV, Hall EM (1988) Job strain, workplace social support and cardiovascular disease: A cross-sectional study of a random sample of the Swedish working population. American Journal of Public Health 78: 1336–42.

Johnson J, Hall E, Theorell T (1989) The combined effects of job strain and social isolation on the prevalence and mortality incidence of cardiovascular disease in a random sample of the Swedish male working population. Scandinavian Journal of Work and Environmental Health 15: 271–9.

Johnson JV, Stewart W, Hall EM, Fredlund P, Theorell T (1996) Long-term psychosocial work environment and cardiovascular mortality among Swedish men. American Journal of Public Health 86: 324–31.

Karasek R, Theorell T (1990) Healthy Work: Stress, Productivity, and the Reconstruction of Working Life. New York: Basic Books.

Kasper AS (1999) Overwork: causes and consequences. Abstract. Work Stress and Health. Organization of work in a global economy. Baltimore, MD, March 11–13.

Katz D, Kahn R (1966) Social Psychology of Organizations. New York: Wiley.

Kavelaars A, Knook L, Prakken B, Kuis W, Heijnen CJ (1997) Dysregulation of the interaction between the neuro-endocrine system and the immune system in children with chronic fatigue syndrome. Abstract. 14th World Congress on Psychosomatic Medicine, Psychosomatic Medicine: Towards the year 2000. 31/8-5/9 1997, Cairns, Australia.

King R (1999) Secondary traumatic stress among Australian mental health case managers. Abstract. Work Stress and Health. Organization of work in a global economy. Baltimore, MD, March 11–13.

Knutsson A (1989) Shift work and coronary heart disease. Scandinavian Journal of Social Medicine Suppl. 44: 1–15.

Knutsson A, Hallqvist J, Reuterwall C, Theorell T, Åkerstedt T (1999) Shiftwork and myocardial infarction: a case-control study. Occupational and Environmental Medicine 56: 46–50.

La Croix AZ (1984) Occupational exposure to high demand/low control work and coronary heart disease incidence in the Framingham cohort. PhD dissertation, Department of Epidemiology, Chapel Hill, NC: University of North Carolina.

Leymann H (1990) Mobbing and psychological terror at workplace. Violence and Victims 5(2): 119–26.

Marmot M, Theorell T (1988) Social class and cardiovascular disease: The contribution of work. International Journal of Health Services 18: 659–74.

Michelsen D, Licinio JG, Gold PW (1995) Meditation of the stress response by the hypothalamic-pituitary-adrenal axis. In: Friedman MJ, Charney DS, Deutch AY (eds) Neurobiological and Clinical Consequences of Stress. Philadelphia, PA: Lippincott-Raven.

Peter R, Alfredsson L, Hammar N, Siegrist J, Theorell T, Westerholm P (1998) High effort, low reward and cardiovascular risk factors in employed Swedish men and women – baseline results from the WOLF-Study. Journal of Epidemiology and Community Health 52: 540–7.

Reed DM, la Croiz AZ, Karasek RA, Miller D, McLean CA (1989) Occupational strain and the incidence of coronary heart disease. American Journal of Epidemiology 129: 495–502.

Reuterwall C, Hallqvist J, Ahlbom A, de Faire U, Diderichsen F, Hogstedt C, Pershagen G, Theorell T, Wiman B, Wolk A (the SHEEP Study Group) (1999) Higher relative but lower absolute risks of myocardial infarction in women than in men: analysis of some major risk factors in the SHEEP Study Group. Journal of Internal Medicine 246(2):161–74.

Rosmond R (1998) Psychoneuroendocrine aspects on the metabolic syndrome. A population-based study of middle-aged men. Academic thesis. Karolinska Institutet, Stockholm, Sweden.

Rubin RT, Poland RE, Lesser IM, Winston RA, Blodgett AL (1987) Neuroendocrine aspects of primary endogenous depression. I. Cortisol secretory dynamics in patients and matched controls. Archives of General Psychiatry 44(4): 328–36.

Schnall PL, Pieper C, Schwartz JE, Karasek RA, Schlussel Y, Devereux RB, Ganan A, Alderman M, Warren K, Pickering T (1990) The relationship between 'job strain', workplace, diastolic blood pressure and left ventricular mass index.Results of a case-control study. Journal of the American Medical Association 263: 1929–35.

Siegrist J (1996) Adverse health effects of high-effort/low-reward conditions. Journal of Occupational Health and Psychology 1: 27–41.

Siegrist J, Matschinger M, Cremer P, Seidel D (1988) Atherogenic risk in men suffering from occupational stress. Atherosclerosis 69: 211–18.

Statistics Sweden (1997)Living conditions and inequality in Sweden – a 20 year perspective 1975–1995. Report 91. Official Statistics of Sweden.

Theorell T (2000) Working conditions and health. In: Berkman L, Kawachi I (eds) Social Epidemiology. Oxford: Oxford University Press.

Theorell T, Karasek R (1996) Current issues relating to psychosocial job strain and cardiovascular disease research. Journal of Occupational Health Psychology 1: 9–26.

Theorell T, Tsutsumi A, Hallqvist J, Reuterwall C, Hogstedt C, Fredlund P, Emlund N, Johnson JV, the SHEEP Study Group (1998) Decision Latitude, job strain, and myocardial infarction: A study of working men in Stockholm. American Journal of Public Health 88(3): 382–8.

Tsutsumi A, Theorell T, Hallqvist J, Reuterwall C, de Faire U (1999) Association between job characteristics and plasma fibrinogen in a normal working population: a cross sectional analysis in referents of the SHEEP study. Journal of Epidemiology and Community Health 6: 348–54.

Ursin H (1997) Sensitization: A mechanism for somatization and subjective health complaints? Behavioral and Brain Sciences 20(3): 469.

Wamala SP, Wolk A, Schenk-Gustavsson K, Orth-Gomér K (1997) Lipid profile and socioeconomic status in healthy middle aged women in Sweden. Journal of Epidemiology and Community Health 51: 400–7.

Wessley S, Chalder T, Hirsch S, Wallace P, Wright D (1997) The prevalence and morbidity of chronic fatigue and chronic fatigue syndrome: A prospective primary care study. American Journal of Public Health 87: 1449–55.

Wilhelmsen L, Johansson S, Rosengren A, Wallin I, Dotevall A, Lappas T (1997) Risk factors for cardiovascular disease during the period 1985–1995 in Goteborg, Sweden. The GOT-MONICA Project. Journal of Internal Medicine 242(3): 199–211.

Chapter 4
A Theory of Occupational Stress

JOHANNES SIEGRIST

Introduction

For more than 20 years Cary Cooper has contributed to the field of occupational stress in a highly productive and successful way. Most recently, together with Kate Sparks, he has raised a critique against 'general' theories of occupational stress by claiming that situation-specific rather than general models of work–strain relationships need to be developed to account for observed variations in occupational settings (Sparks and Cooper, 1999). Using an approach that incorporates a greater range of variables, the authors identify seven relevant job characteristics: work control, job-intrinsic factors (e.g. work overload, job variety), organizational role (e.g. role conflict, role ambiguity), relationships with others (e.g. social support at work), career and achievement (e.g. over- or under-promotion), organizational climate (e.g. lack of communication), and home/work interface (Sparks and Cooper, 1999). According to them, 'models which encompass a larger range of variables may explain more of the variance in strain outcomes, giving a clearer picture of the work-strain relationship, and ultimately providing the framework for a more effective stress management intervention' (Sparks and Cooper, 1999, p. 220).

At first glance, this critique seems well justified. Nevertheless, the present contribution challenges the proposition of 'situation-specific' models of occupational stress and argues in favour of a 'general' theory. I do hope that this type of theoretical discussion may contribute to future developments of the field, especially so, if rooted in an atmosphere of mutual respect and friendship. The first part of this contribution summarizes core statements of one example of such a 'general' theory of occupational stress, termed effort–reward imbalance. This approach has been developed and, in part, tested by the author and his group for several years (for a recent summary see Siegrist, 1998). In the next part the main empirical evidence in support of this approach is summarized; in a final section, future developments and policy implications are discussed.

To start, it is useful to clarify what I understand by the term 'theory', and to define some basic terms that are frequently used in the field of occupational stress research. A theory may best be understood as a set of general statements that offer an explanation (or prediction) of an association of phenomena that has not yet been understood. The wider a range of phenomena that are explained by one single theory, the more advanced and powerful a theory, and the more economic its formulations (Popper, 1959). Theoretical statements are based on risky assumptions, i.e. assumptions that deviate from those predictions that are based on common sense or on what is already known. Therefore, a (preliminary) confirmation of a theory has the potential of creating new knowledge. However, a theory should be designed in a way that excludes certain alternative explanations, and it should be measured in a way that permits a test of its validity, i.e. its falsification or its confirmation. It is admitted that the development of theories in the social and behavioural sciences is a difficult task, given the complexities, the dynamics and the variability of human behaviour. Therefore, it has been proposed that developing theories in these fields may be restricted to so-called middle-range theories whose level of generalization is limited in time and space (Merton, 1957). Along these lines, the following theory is restricted to the prediction of physical and mental health produced by a specific psychosocial environment at work that can be identified quite frequently in economically advanced societies.

Before introducing the model of effort–reward imbalance, the following basic terms are defined. Much critique has been raised against the ambiguity of the term 'stress'. To avoid this, the following definitions are suggested: 'stressor' is defined as an environmental demand or threat that taxes or exceeds a person's ability to meet the challenge; 'strain' is defined as the person's response to such a situation in psychological and physiological terms. Psychological responses include negative emotions (e.g. anger, frustration, anxiety, helplessness) whereas physiological responses concern the activation of the autonomic nervous system and related neuro-hormonal and immune reactions. Stressors, in particular novel or dangerous ones, are appraised and evaluated by the person, and as long as there is some perception of agency on the part of the exposed person, efforts are mobilized to reverse the threat or to meet the demands. Such efforts are termed 'coping', and they occur at the behavioural (even interpersonal), cognitive, affective and motivational level. Clearly, when judging strain, the quality and intensity of a stressor as well as the duration of exposure have to be taken into account, as well as individual differences in coping and in vulnerability to strain reactions. Recent research indicates that only part of human strain reactions are subject to conscious information processing whereas a large amount by-passes awareness. The term 'stressful experience' is introduced to delineate that part of affective processing that reaches consciousness (see below). Stressful experience at work is often attributed to adverse working conditions by exposed people themselves. While they usually refer to some common sense notions of stress it is crucial to note that these attributions differ from the explanatory constructs of stressful experience at work that have been identified by science.

The model of effort–reward imbalance

Research on psychosocial occupational stress differs from traditional biomedical occupational health research in that stressors cannot be identified by direct physical or chemical measurements. Rather, theoretical models are needed to analyse the particular nature of the psychosocial work environment. Ideally, a theoretical model of stressor–strain relationships at work should encompass a wide variety of different occupations. Moreover, it should account for the fact that people exposed to stressors tend to avoid or reduce their strain reactions, at least as far as these reach their awareness, to minimize potential adverse consequences. In other words, a theoretical model should specify the conditions under which strain reactions are likely to become chronically recurrent. Finally, a theoretical model should indicate to what extent strain reactions are elicited by specific situational (extrinsic) conditions or by specific intrinsic characteristics of the person exposed to these conditions.

The model of effort–reward imbalance puts its selective emphasis on the fact that the work role defines a crucial link between self-regulatory needs of a person such as self-esteem and self-efficacy, and the social opportunity structure. Conferment of occupational status is associated with recurrent options of contributing and performing, of being rewarded or esteemed, and of belonging to some significant group (work colleagues). Yet, these potentially beneficial effects are contingent on a basic prerequisite of exchange in social life, that is, reciprocity. Effort at work is spent as part of a socially organized exchange process to which society at large contributes in terms of rewards. Rewards are distributed by three transmitter systems: money, esteem, and career opportunities including job security. The model claims that lack of reciprocity between 'costs' and 'gains' (i.e. high cost/low gain conditions) elicits sustained strain reactions at the emotional and physiological levels. For instance, having a demanding, but unstable job, achieving at a high level without being offered any promotion prospects, are examples of high cost/low gain conditions at work. In terms of current developments of the labour market in a global economy, the emphasis on occupational rewards including job security reflects the growing importance of fragmented job careers, of job instability, under-employment, redundancy and forced occupational mobility including their financial consequences. The model of effort–reward imbalance applies to a wide range of occupational settings, most markedly to groups that suffer from a growing segmentation of the labour market, to groups exposed to structural unemployment and rapid socio-economic change. Effort–reward imbalance is frequent among service occupations and professions, in particular the ones dealing with client interactions.

But how can we be sure that these widely prevalent high cost/low gain conditions at work elicit chronically recurrent strain reactions? In terms of the expectancy value theory of motivation, it is likely that workers exposed to high effort/low reward conditions give up this state or, if that is not feasible, reduce their efforts to minimize negative outcome (Schönpflug and Batman, 1989), at least as far as strain is consciously appraised as stressful experience. Contrary to this theory the model of effort–reward imbalance predicts continued high effort and, thus, chronically stressful experience, under the following conditions: (a) lack of alternative choice in the labour market may prevent people from giving up even unfavourable jobs, as the anticipated costs of this engagement (e.g. the risk of being laid off or of facing downward mobility) outweigh costs of accepting inadequate benefits; (b) unfair job arrangements may be accepted for a certain period of one's occupational trajectory for strategic reasons; by doing so employees tend to improve their chances for career promotion and related rewards at a later stage; (c) a specific personal pattern of coping with demands and of eliciting rewards characterized by overcommitment may prevent people from accurately assessing cost–gain relations. 'Overcommitment' defines a set of attitudes, behaviours and emotions reflecting excessive striving in combination with a strong desire of being approved and esteemed. People characterized by overcommitment are exaggerating their efforts beyond levels usually considered appropriate. There is good reason to suggest that excessive efforts result from perceptual distortion (e.g. underestimation of challenge) which in turn may be triggered by an underlying motivation of experiencing recurrent esteem and approval (Siegrist, 1996). This latter argument points to the third point raised above: it defines a person-specific component of the model ('overcommitment') in addition to the situation-specific components of high extrinsic effort and low reward.

A graphic representation of the model is given in Figure 4.1. The figure highlights the stress-theoretical relevance of an imbalance between high effort and low reward at work that may be due to the fact that violations of the social norm of reciprocity conflict with essential expectancies that are often taken for granted. These expectancies are assumed to be 'imprinted' in humans' evolutionary brain structures as a basic grammar of social exchange, the grammar of reciprocity and fairness (Cosmides and Tooby, 1992). Moreover, the figure highlights the three dimensions of reward experience or future reward expectancy distinguished by the model: money, esteem and career opportunities. It is obvious that the three types of reward relate to different experiences. While salaries and wages as well as promotion prospects and job security are linked to more distant organizational and macro-economic labour market conditions, esteem reward is related to the more proximate interpersonal experience of favourable social exchange. In the model, no a priori specification is made concerning health effects of the different types of reward. Rather, it is the mismatch between 'high costs' spent and 'low gain' received which matters most. Finally, the figure illustrates an explicit distinction made between extrinsic, situational and intrinsic, personal components of effort–reward imbalance. It assumes that a combination of both sources of information provides a more accurate estimate of the total amount of strain attributable to work than a restriction to one of these two sources (either situational or personal). However, as situational and personal components are clearly distinct at the conceptual and at the operational level, the relative contribution of each component to an explanation of adverse health can be assessed in quantitative terms.

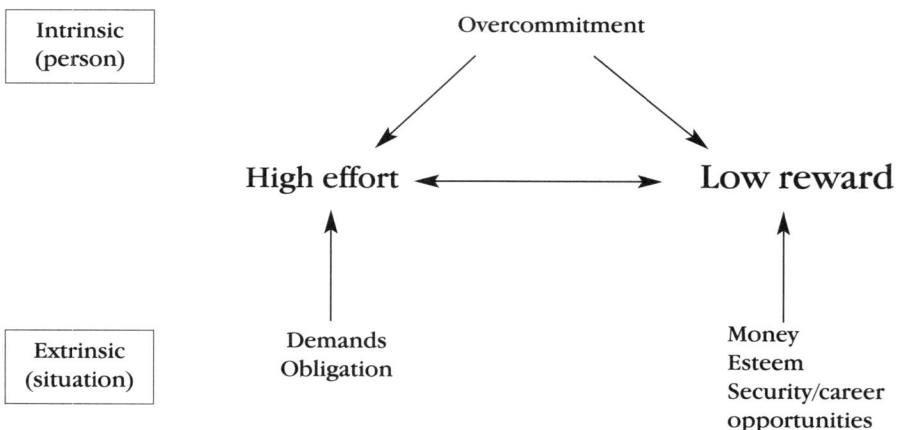

Figure 4.1: The Effort-Reward Imbalance Model.

Empirical support

Some twelve independent investigations have tested the effort–reward imbalance model so far. Most studies were concerned with coronary heart disease or cardiovascular risk factors as outcome measures, and four of these were prospective or semi-prospective epidemiological investigations (Bosma, Peter, Siegrist and Marmot, 1998; Joksimovic, Siegrist, Meyer-Hammer, Peter, Franke, Klimek, Heintzen and Strauer, 2000, Siegrist, Peter, Junge, Cremer and Seidel, 1990a; Siegrist, Bernhardt, Feng and Schettler, 1990b). Three further studies related to cardiovascular health were cross-sectional (Peter, Alfredsson, Hammar, Siegrist, Theorell and Westerholm, 1998; Peter, Siegrist, Hallqvist, Reuterwall, Theorell and the SHEEP Study Group, 1999; Siegrist, Peter, Cremer and Seidel, 1997), one investigation concerned gastrointestinal symptoms (Rothenbacher, Peter, Bode, Adler and Brenner, 1998). Finally, three additional cross-sectional studies were related to indicators of subjective health, e.g. perceived symptoms, burnout (Bakker, Killmer, Siegrist and Schaufeli, 2000; Jonge, Bosma, Peter and Siegrist, 2000; Peter and Siegrist, 1997), and one investigation tested the effects of a theory-based intervention on level of well-being and overcommitment (Aust, Peter and Siegrist, 1997). It should be mentioned that in one prospective investigation, the British Whitehall II Study, identical measures of effort–reward imbalance were related to different health outcomes, that is firstly to new reports of coronary heart disease (Bosma et al., 1998), secondly to new reports of mild psychiatric disorders (Stansfeld, Fuhrer, Shipley and Marmot, 1999), and thirdly to level of functioning (Stansfeld, Bosma, Hemingway and Marmot, 1998). It is important to stress that some studies used proxy measures rather than the original questions designed by the authors (Siegrist, 1996; Peter et al., 1998), but in almost all cases, a co-manifestation of indicators of high effort *and* indicators of low reward was observed to produce strongest health effects, as postulated by theory. More recently, a highly formalized rigorous measurement approach has been proposed to define the critical threshold of effort–reward imbalance in quantitative terms (Peter et al., 1998).

A brief summary of some major findings is given here. First, based on current evidence and according to the occupation under study, between 10% and 40% of the workforce suffer from some degree of effort–reward imbalance at work, and at least a third of them are characterized by sustained intense strain reactions following exposure to effort–reward imbalance. In general, these strain reactions are more frequent among lower socio-economic groups, thus pointing to a possible contribution of the model towards explaining part of the increased health burden observed among middle-aged economically active populations with lower socio-economic status (Marmot, Siegrist, Theorell and Feeney, 1999).

Secondly, with regard to future incident coronary heart disease, effort–reward imbalance at work was associated with a 2.7- to 6.1-fold elevated relative risk compared to those who were free from chronic strain at work. This excess risk could not be explained by established biomedical and behavioural risk factors as these variables were taken into account in multivariate statistical analysis. Thus, the psychosocial work environment as measured by this model is associated with at least a moderate relative risk of incident coronary heart disease that is independent of established biomedical and behavioural cardiovascular risk factors. However, restricting the analysis to this association would result in an underestimation of the total burden on cardiovascular health produced by adverse psychosocial work conditions. This is due to the fact that chronic psychosocial strain at work is also associated with relevant cardiovascular risk factors, e.g. high blood pressure (hypertension), high levels of blood lipids, or a co-manifestation of these two risk factors (Peter and Siegrist, 1997; Peter et al., 1998; Siegrist et al., 1997; Siegrist, Peter, Georg, Cremer and Seidel, 1991). These findings demonstrate that the explanatory power of the model goes beyond disease manifestation by enabling a more comprehensive definition of people at risk at an earlier stage of disease development.

Thirdly, effort–reward imbalance was associated with moderately elevated risks of impaired physical, mental and social functioning (odds ratios ranging from 1.40 to 1.78 in men and from 1.81 to 2.33 in women; Stansfeld et al., 1998) and with moderately elevated risks of newly reported mild psychiatric disorders (odds ratios ranging from 1.67 in women to 2.57 in men; Stansfeld et al., 1999) in the Whitehall II study. Additional evidence along these lines come from cross-sectional studies, confirming that the explanatory power of the model covers more than some specific physical diseases and biomedical risk factors; it also includes aspects of mental health and of physical, mental and social functioning. In other words, an estimation of the total burden of health produced by occupational strain, as measured by this model, far exceeds the amount identified by studies that focus on a single outcome measure. This conclusion is relevant both in theoretical and in practical terms.

In theoretical terms, findings seem to fit into the stress-physiological concept of 'generalized illness susceptibility' (Cassel, 1976). According to this concept, neuro-hormonal and immune reactions following central nervous system-induced strain have the potential of triggering and precipitating those pathophysiological processes that are developed already to some extent in an individual, either by genetic predisposition or by behaviourally induced risk factors. In other words, strain is not assumed to account for the total chain of events in a pathophysiological process of disease development. Rather, it selectively reinforces and critically advances preclinical stages into overt clinical manifestation of a disease although this process may often take years rather than months or weeks.

This view has found indirect support in a series of results relating effort–reward imbalance at work to cardiovascular risk and disease. Although, in statistical terms, effort–reward imbalance predicted incident coronary heart disease after adjusting for powerful confounders, such as established cardiovascular risk factors, expected probabilities calculated on the basis of the beta-coefficients of logistic regression models revealed that statistical prediction was highest when the conventional risk factors and the newly detected psychosocial risk factor were combined. To give an example: based on a respective logistic regression model the expected probability of an incident coronary event over a mean of 6.5 years was 85.3% in blue-collar men who suffered from four conventional biomedical or behavioural cardiovascular risk factors *and* effort–reward imbalance simultaneously, but was less than half if these conventional risk factors were present without exposure to effort–reward imbalance (Siegrist et al., 1990a). Similar findings were observed when estimating the risk of a co-manifestation of hypertension and atherogenic lipids (Siegrist, 1996) or when predicting a coronary restenosis following percutaneous transluminal coronary angioplasty in cardiac patients (Joksimovic et al., 1999). Experimental animal stress research provides additional evidence along these lines (Kaplan and Manuck, 1994).

Although it is admitted that effort–reward imbalance, according to the concept of illness susceptibility, can be related to a range of health outcomes, this application may be most successful with regard to cardio-vascular health. This is the case because of the specific nature of emotional and physiological responses that are evoked by effort–reward experience. This response can be described as striving without success, as continued activation without relaxation, as energy mobilization without recovery, or as joyless struggling paralleled by feelings of disappointment and frustration. There is stress-physiological evidence from animal research that a synergistic activation of two stress axes occurs under such conditions, the hypothalamic-pituitary-adrenocortical and the sympatho-adreno-medullary stress axis. This synergism was shown to produce particularly strong pathophysiological effects on the cardiovascular system (Henry, 1992; Kaplan and Manuck, 1994).

Subjective health is another important outcome measure. Although it is not clear whether physiological processes are involved in mediating the association of work-related strain with subjective health, several studies indicate that this latter variable in itself is a powerful predictor of morbidity and mortality (Idler and Benyamini, 1997). In view of this observation the finding of moderate or even strong associations of effort–reward imbalance with subjective health (odds ratios ranging from 1.4 to 4.4 in several studies) deserves further attention. Before discussing some obvious policy implications of the results summarized in this section, the next section starts with a discussion of some open issues and future developments related to this example of a 'general', as opposed to a situation-specific, theoretical approach.

Future directions and policy implications

One obvious open issue with respect to the effort–reward imbalance model concerns its difference from, or overlap with, a highly influential, well established alternative 'general' theory of occupational stress, the demand–control model (Karasek, 1979; Karasek and Theorell, 1990). This approach claims that a high level of psychological demands combined with a low level of decision latitude (low level of decision authority and low level of skill utilization) elicits sustained strain and thus contributes to illness susceptibility. A large number of investigations confirmed that people working in jobs that are defined by high demands and low task control are at moderately elevated risk of chronic disease manifestation (Theorell and Karasek, 1996). Most convincingly so far, this was shown for cardiovascular disease (Marmot et al., 1999). Although the two models overlap with regard to the dimension of demands/extrinsic efforts they nevertheless differ in the following regards. First, the demand–control model has been introduced and measured as a concept that is restricted to the situational aspects of the psychosocial work environment whereas the effort–reward imbalance model includes both situational and person characteristics (see above and Figure 4.1). Secondly, the demand–control model offers a broader approach as its bi-directional conceptualization includes a stress dimension with relevance to health and a skill dimension with relevance to personal growth and development: Active jobs characterized by high demands and a high degree of control and opportunities for skill utilization promote personal growth and feelings of mastery (Karasek and Theorell, 1990). In this regard, the model presented here is more narrowly focused on the bio-psycho-social determinants of health and well-being. Thirdly, components of the effort–reward imbalance model (salaries, career opportunities/job security) are linked to more distant macro-economic labour market conditions while the demand–control model's major focus is on workplace characteristics. In stress-theoretical terms, Karasek's approach is rooted in the stress-theoretical paradigm of personal control that has attracted wide attention from several scientific disciplines (see e.g. Skinner, 1996; Spector, 1998; Steptoe and Appels, 1989). However, the model of high cost/low gain conditions fits better with the stress-theoretical paradigm of social reward that emphasizes the powerful regulatory role of a particular brain system implicated in motivation, reinforcement and reward (Henry and Stephens, 1977). Finally, the two different stress-theoretical orientations have different implications for policy: whereas the control paradigm points to the structure of power, division of labour and democracy at work, the reward paradigm addresses the issue of distributive justice and fairness.

As conditions of low personal control and low social reward may often cumulate in a person's work and life setting it seems promising to study the separate and combined effects on health produced by both models discussed so far. In fact, preliminary evidence from a large-scale study indicates that combined effects on cardiovascular health are considerably stronger compared to the separate effects of each model (Peter et al., 2000). Moreover, in the Whitehall II study (Bosma et al., 1998) it was found that the demand–control model (the control dimension only) and the effort–reward imbalance model were equally strong independent predictors of reports of incident coronary heart disease when appropriate statistical controls were performed.

An additional open issue concerns a more appropriate conceptualization of the cumulative effects of experiencing effort–reward imbalance in a lifetime perspective. For instance, elderly workers may be unable to maintain continuously high efforts on their job because after years or decades of exposure their resources are exhausted. If this situation is not reflected in compensatory wage differentials work strain is expect to be much worse than if experienced by a young worker in a comparable situation. Similarly, the personal coping pattern termed overcommitment needs to be conceptualized in a more dynamic, time sensitive way as it is unlikely that it operates as a stable personality characteristic.

If the general assumption of the theoretical approach presented above holds true we can expect similar adverse effects on health produced by conditions of effort–reward imbalance that are generated in other core social roles (e.g. partnership, family). In fact, an extension of the model beyond work is now under way. In this context, the interface between work and non-work settings, in particular work and family, needs to be explored more thoroughly. Three terms have been suggested in this context: 'spillover', 'compensation' and 'cumulation'. Cumulation describes the fact that critical components of strain experience, such as lack of control or effort–reward imbalance, generalize across different life domains. 'Spillover' refers to the transfer of experience and behaviour from one domain to the other, whereas 'compensation' represents efforts to reduce strain in one domain by improving satisfaction in another domain.

At the beginning it was stated that a general theory of occupational strain with relevance to health can be applied to a wide range of different occupations. This wide application is due to the fact that the core terms of such a theory are rather global. By selectively focusing on an analytical frame defined by these terms a theory is intended to identify the 'toxic' components within a broader stream of work-related experience. By 'toxic' I mean those components that elicit sustained intense strain reactions with adverse long-term consequences for health. The theory maintains that, once a stressful working context meets the criteria of a high cost/low gain condition, roughly the same processes of emotional and physiological activation are observed in subjects who are exposed to these conditions. Clearly, their ultimate health consequences may vary according to the duration and intensity of the stressor and according to individual coping abilities and illness susceptibility. Such a selective, general theory runs the risk of being falsified as its predictions are rather specific ones. For instance, effort–reward theory specifically predicts that the simultaneous manifestation of conditions of high effort and of low reward at work is required to produce the intensity of strain reactions that is needed to disturb bodily functions (Weiner, 1992).

Perhaps, the ultimate justification of a general theory of occupational stress, and the most obvious difference to any situation-specific modelling of occupational stress, can be found in the basic distinction between the terms 'strain' and 'stressful experience' (see above). As long as these terms are used synonymously, that is, as long as a cognitive theory of stress is applied, it is mandatory to explore a wide range of potential work-related stressors by referring to the subject's own appraisal. Cognitive stress theory claims that those stressors only evoke stressful experience or strain that are consciously appraised by exposed people (Lazarus, 1991). As long as occupational stress research is guided by this approach, it is quite convincing to operate with a comprehensive, broad list of potential stressors from which subjects can select the appropriate ones. Yet, if we admit that affective processing is different from conscious computational processing and that only part of the flow of information that matters for affective processing (= strain) is subject to conscious appraisal (= stressful experience), we are then entitled to apply a more restrictive, theory-based set of indicators to identify the 'toxic' components mentioned. In fact, the concept and measurement of effort–reward imbalance is rooted in a theory of emotional stress that incorporates two different principles of information processing, affective processing and conscious computational processing: 'In contrast to computational processing we have no control over the way in which emotional aspects of the information are processed. These processes are encapsulated and are largely unconscious. Only the results of this processing reach our consciousness. We may even feel anxious although we do not know why' (Gaillard and Wientjes, 1993, p. 268; see also LeDoux, 1996). In this perspective, negative affect associated with experience of effort–reward imbalance at work may not necessarily be subject to conscious appraisal, especially as it is a chronically recurrent everyday experience. Based on this assumption, the measurement of effort–reward imbalance at work, although relying on the respondent's subjective judgements, requires the computation of a summary measure according to a predefined algorithm (Peter et al., 1998). It is assumed that a theory-based reconstruction of strain based on this algorithm offers a more comprehensive assessment of both types of information processing. Again, this is an issue of further in-depth inquiry and research.

What are the policy implications of this new information? First, it is possible to identify dimensions of work-related strain in a wide range of occupations using standardized, well-tested questionnaires. These questionnaires measuring the effort–reward imbalance model, as well as the demand–control model, are now available in a number of languages internationally. Thus, further scientific evidence on associations between adverse psychosocial work environments and health indicators in working populations can be obtained. Secondly, it is possible to apply these measures beyond scientific purpose to evaluate the amount of work-related strain, e.g. in an enterprise or in a specific occupational group. With the help of a computerized statistical program respective information can be fed back to those concerned, e.g. to serve as a basis of monitoring activities or of a stress prevention programme. A third possible application concerns legal procedures and compensation claims regarding the afflictions of work life on health. Here again, quantified, evidence-based information can be useful to support decision-making processes. Yet, it should be mentioned that quantitative evidence on the proportion of a health risk that is attributable to work-related strain is confined to the level of populations, not individuals. Thus, the 'aetiological fraction' that is attributable to adverse work conditions can hardly be transferred to the individual case, for instance in the context of justification of a compensation claim (Rockhill, Newman and Weinberg, 1998).

Probably the most significant policy implication of the information provided above concerns the design and implementation of worksite stress prevention and health promotion programmes. Both approaches, the effort–reward imbalance and the demand–control model, offer specific suggestions in this respect. Whereas propositions derived from the demand–control model are related to measures of job redesign, job enlargement, job enrichment, skill training and enhanced participation (Karasek, 1992), the current model's focus is on adequate terms of exchange between efforts and rewards. Examples of such measures include the development of compensatory wage systems, the provision of models of gain sharing and the strengthening of non-monetary gratifications. Moreover, ways of improving promotional opportunities and job security need to be explored. Supplementary measures are interpersonal training and social skills development, in particular leadership behaviour. For instance, one recent stress management intervention based on the model was successfully applied in a group of highly strained inner-city bus drivers (Aust et al., 1997). In conclusion, multiple efforts will be needed at both the scientific and policy level to promote healthy work, especially in a rapidly changing occupational world.

References

Aust B, Peter R, Siegrist J (1997) Stress management in bus drivers: a pilot study based on the model of effort–reward imbalance. International Journal of Stress Management 4: 297–305.

Bakker AB, Killmer CH, Siegrist J, Schaufeli WB (2000) Effort–reward imbalance and burnout among nurses. Journal of Advanced Nursing 31: 884–91.

Bosma H, Peter R, Siegrist J, Marmot MG (1998) Two alternative job stress models and the risk of coronary heart disease. American Journal of Public Health 88: 68–74.

Cassel J (1976) The contribution of the social environment to host resistance. American Journal of Epidemiology 104: 107–14.

Cosmides L, Tooby J (1992) Cognitive adaptations for social exchange. In: Barkow JH, Cosmides, Tooby J (eds) The Adapted Mind: Evolutionary Psychology and the Generation of Culture, pp. 163–228. New York: Oxford University Press.

Gaillard AWK, Wientjes CJE (1993) A framework for the evaluation of work stress by physiological reactivity. In: La Ferla F, Levi L (eds) A Healthier Work Environment, pp. 266–82. Copenhagen: World Health Organization.

Henry JP (1992) Biological basis of the stress response. Integrative Physiological and Behavioral Science 27: 66–83.

Henry JP, Stephens PM (1977) Stress, Health, and the Social Environment. Berlin: Springer.

Idler EL, Benyamini Y (1997) Self-rated health and mortality: a review of twenty seven community studies. Journal of Health and Social Behavior 38: 21–37.

Joksimovic L, Siegrist J, Meyer-Hammer M, Peter R, Franke B., Klimek WJ, Heintzen M, Strauer BE (2000) Overcommitment predicts restenosis after successful coronary angioplasty in cardiac patients. International Journal of Behavioral Medicine 6: 356–369.

Jonge J, Bosma H, Peter R, Siegrist J (2000) Job strain effort–reward imbalance and employee well-being: a large-scale cross-sectional study. Social Science and Medicine 50(9): 131–27

Kaplan JR, Manuck SB (1994) Anti-atherogenic effects of beta-adrenergic blocking agents: theoretical, experimental, and epidemiological consideration. American Heart Journal 128: 1316–29.

Karasek R (1979) Job demands, job decision latitude, and mental strain: implications for a job re-design. Administration Science Quarterly 24: 285–307.

Karasek R (1992) Stress prevention through work reorganization: a summary of 19 international case studies. Conditions of Work Digest 11: 23–41.

Karasek R, Theorell T (1990) Healthy Work: Stress, Productivity, and the Reconstruction of Working Life. New York: Basic Books.

Lazarus RS (1991) Emotion and Adaptation. New York: Oxford University Press.

LeDoux J (1996) The Emotional Brain. New York: Simon & Schuster.

Marmot M, Siegrist J, Theorell T, Feeney A (1999) Health and the psychosocial environment at work. pp. 105–131. In: Marmot MG, Wilkinson R (eds) Social Determinants of Health. Oxford: Oxford University Press.

Merton RK (1957) Social Theory and Social Structure. New York: Free Press.

Peter R, Siegrist J (1997) Chronic work stress, sickness absence, and hypertension in middle managers: general or specific sociological explanations? Social Science and Medicine 45: 1111–20.

Peter R, Alfredsson L, Hammar N, Siegrist J, Theorell T, Westerholm P (1998) High effort, low reward, and cardiovascular risk factors in employed Swedish men and women: baseline results from the WOLF Study. Journal of Epidemiology and Community Health 52: 540–7.

Peter R, Siegrist J, Hallqvist J, Reuterwall C, Theorell T, the SHEEP Study Group (2000) Psychosocial work environment and myocardial infarction: improving risk estimation by combining two alternative job stress models in the SHEEP Study (submitted).

Popper KR (1959) The Logic of Scientific Discovery. London: Routledge.

Rockhill B, Newman B, Weinberg C (1998) Use and misuse of population attributable fractions. American Journal of Public Health 88: 15–19.

Rothenbacher D, Peter R, Bode G, Adler G, Brenner H (1998) Dyspepsia in relation to Helicobacter pylori infection and psychosocial work stress in white collar employees. American Journal of Gastroenterology 93: 1443–9.

Schönpflug W, Batman W (1989) The costs and benefits of coping. In: Fisher S, Reason J (eds) Handbook of Stress, Cognition and Health, pp. 699–714. Chichester: Wiley.

Siegrist J (1996) Adverse health effects of high effort – low reward conditions at work. Journal of Occupational Health Psychology 1: 27–43.

Siegrist J (1998) Adverse health effects of effort–reward imbalance at work. In: Cooper CL (ed.) Theories of Organizational Stress, pp. 190–204. Oxford: Oxford University Press.

Siegrist J, Peter R, Junge A, Cremer P, Seidel D (1990a) Low status control, high effort at work and ischaemic heart disease: prospective evidence from blue collar men. Social Science and Medicine 31: 1127–34.

Siegrist J, Bernhardt R, Feng Z, Schettler G (1990b) Socioeconomic differences in cardiovascular risk factors in China. International Journal of Epidemiology 19: 905–10.

Siegrist J, Peter R, Georg W, Cremer P, Seidel D (1991) Psychosocial and behavioral characteristics of hypertensive men with elevated atherogenic lipids. Atherosclerosis 86: 211–18.

Siegrist J, Peter R, Cremer P, Seidel D (1997) Chronic work stress is associated with atherogenic lipids and elevated fibrinogen in middle aged men. Journal of Internal Medicine 242: 149–56.

Skinner EA (1996) A guide to constructs of control. Journal of Personality and Social Psychology 71: 549–70.

Sparks K, Cooper CL (1999) Occupational differences in the work-strain relationship: Towards the use of situation-specific models. Journal of Occupational and Organizational Psychology 72: 219–29.

Spector PE (1998) A control theory of the job stress process. In: Cooper CL (ed.) Theories of Organizational Stress, pp. 153–69. Oxford: Oxford University Press.

Stansfeld S, Bosma H, Hemingway H, Marmot M (1998) Psychosocial work characteristics and social support as predictors of SF-36 functioning: the Whitehall II Study. Psychosomatic Medicine 60: 247–55.

Stansfeld S, Fuhrer R, Shipley MJ, Marmot M (1999) Work characteristics predict psychiatric disorders: prospective results from the Whitehall II Study. Occupational and Environmental Medicine 56(5): 302–7.

Steptoe A, Appels A (eds) (1989) Stress, Personal Control and Health. Chichester: Wiley.

Theorell T, Karasek R (1996) Current issues relating to psychosocial job strain and cardiovascular disease research. Journal of Occupational Health Psychology 1: 9–26.

Weiner H (1992) Perturbing the Organism: The Biology of Stressful Experience. Chicago, IL: Chicago University Press.

Chapter 5
The Stressful Effects of Mergers and Acquisitions

SUSAN CARTWRIGHT AND SHEILA PANCHAL

Introduction

The incidence of mergers and acquisitions (M&As) has continued to increase significantly during the last decade and has become a growing area of research attention, both domestically and internationally (Gertsen, Søderberg and Torpe, 1998). Forecasts suggest that it is unlikely that the frequency of M&As will diminish in the near future (Balzer, 1997). Although M&As are popular business strategies they are highly susceptible to failure. Cartwright and Cooper (1996) have presented research evidence that suggests that no more than 50% of M&As achieve the standards of success initially anticipated. Other sources (e.g. Marks, 1988; International Survey Research (ISR), 1999) have estimated even higher failure rates.

There is a growing body of research which has attributed poor M&A performance to human factors (Holbeche, 1998). Walsh (1999) reports in a study of 179 merged or acquired organizations that the majority experienced employee relations problems. Only 30% had integrated their workforce smoothly and just 34% had been able to maintain employee morale throughout the transition. Insensitive management, poor communication and cultural incompatibility are commonly cited reasons for merger failure (Hall and Norburn, 1987; Cartwright and Cooper, 1996).

This chapter presents a review of the current literature which has investigated individual responses to mergers, acquisitions and other large-scale organizational change events within a stress-coping framework. Sources and outcomes of M&A stress will be discussed as well as factors which moderate the stress–strain relationship. Whilst mergers and acquisitions are legally different transactions, in the main the two terms will be treated synonymously on the basis that in practice mergers are rarely marriages of equals and there is invariably a 'buyer' and a 'bought' (Cartwright and Cooper, 2000).

Individual responses to organizational change

The relationship between change and stress has been extensively researched (Ashford, 1988; Nelson, Cooper and Jackson, 1995) within the context of organizational restructuring. However, the unexpected and non-routine nature of the M&A event is considered to set it apart from other forms of organizational change and to increase its stressful potential. According to Schweiger and Ivancevich (1985) M&As are particularly stressful because individuals are unlikely to have developed an effective repertoire of coping strategies to deal with the situation. Therefore, the individual response can be extreme, as in the reported case (McManus and Hergert, 1988) of an employee who was expecting a promotion but instead was made redundant following the acquisition of his company, and subsequently committed suicide. A merger is a significant life event and in terms of the Social Readjustment Rating Scale (Holmes and Rahe, 1967) has been equated with the stress of gaining a new family member or becoming bankrupt.

Models of responses to change

Stage models

Hunsaker and Coombs (1988) interviewed 70 employees one year post-merger, and developed a nine stage model of emotional response to merger. This has been named the 'Merger Emotions Syndrome' (Figure 5.1) and is based on the Kubler-Ross (1969) bereavement framework. The starting point is denial at news of the merger, and follows through a set of negative emotions to reach acceptance as the turning point to positive emotions, culminating in enjoyment/excitement and commitment to the new situation. Individual differences are also considered in terms of emotional outlook. People are classified as hot, cool, warm, pessimistic and optimistic, and it is speculated these differences affect progress through the stages. For example, optimists are more likely to proceed directly into acceptance and experience positive emotions whereas pessimists may experience negative emotions for longer, and could become fixated in the negative side of the syndrome.

Stage theories such as that of Hunsaker and Coombs (1988) are intuitively appealing and have been cited as useful tools for management training concerning understanding and coping with change (Makin, Cooper and Cox, 1996). They are, however, largely descriptive in nature

News of the Merger   Commitment to the New Situation

Denial   Enjoyment

Fear   Liking

Anger   Interest

Sadness   Relief

Acceptance

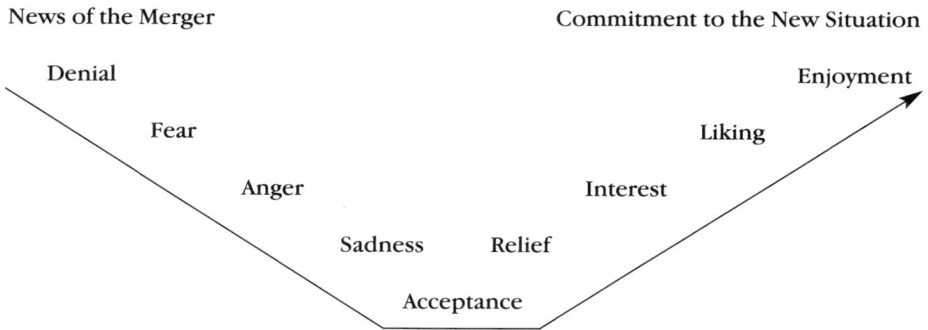

Figure 5.1: The 'Merger Emotions Syndrome' (Hunsaker and Coombs, 1988).

and there is little empirical evidence to support stage theories, possibly because they are difficult to test in terms of time-scales and the effects of individual differences in the rates of change (Kessler, Price and Wortman, 1985). McManus and Herbert (1988) also assert that although stage theories provide a useful basis for comprehending the change response, the psychology of powerlessness needs to be understood more profoundly.

Stress models

An alternative framework offered to understand individual response to change is the stress/coping perspective. Cummings and Cooper (1979) define stress as, 'any force that puts a psychological or physical factor beyond its range of stability, producing a strain within the individual'.

Merger-related stressors can be encapsulated with a general occupational stress model (Cooper and Marshall, 1978). The model (Figure 5.2) conceptualizes stress as emanating from six major sources: job factors, role in organization, work relationships, career development, organizational structure/culture and home–work interface. As M&As are likely to impact upon all six stressor categories, this clearly enhances their stressful potential. The original Cooper and Marshall (1978) model has been criticized for overemphasis on the individual, implying stress is a personal rather than organizational problem (Clarke, Chandler and Barry, 1996). Nevertheless, it provides a useful structure for understanding occupational stress in the context of merger.

Ivancevich, Schweiger and Power (1987) describe a transactional model of merger stress as a complicated process triggered by two major factors, the nature of the merger events and individual characteristics. These factors interact to produce a cognitive appraisal of the merger situation and its anticipated changes. If the changes are appraised as presenting

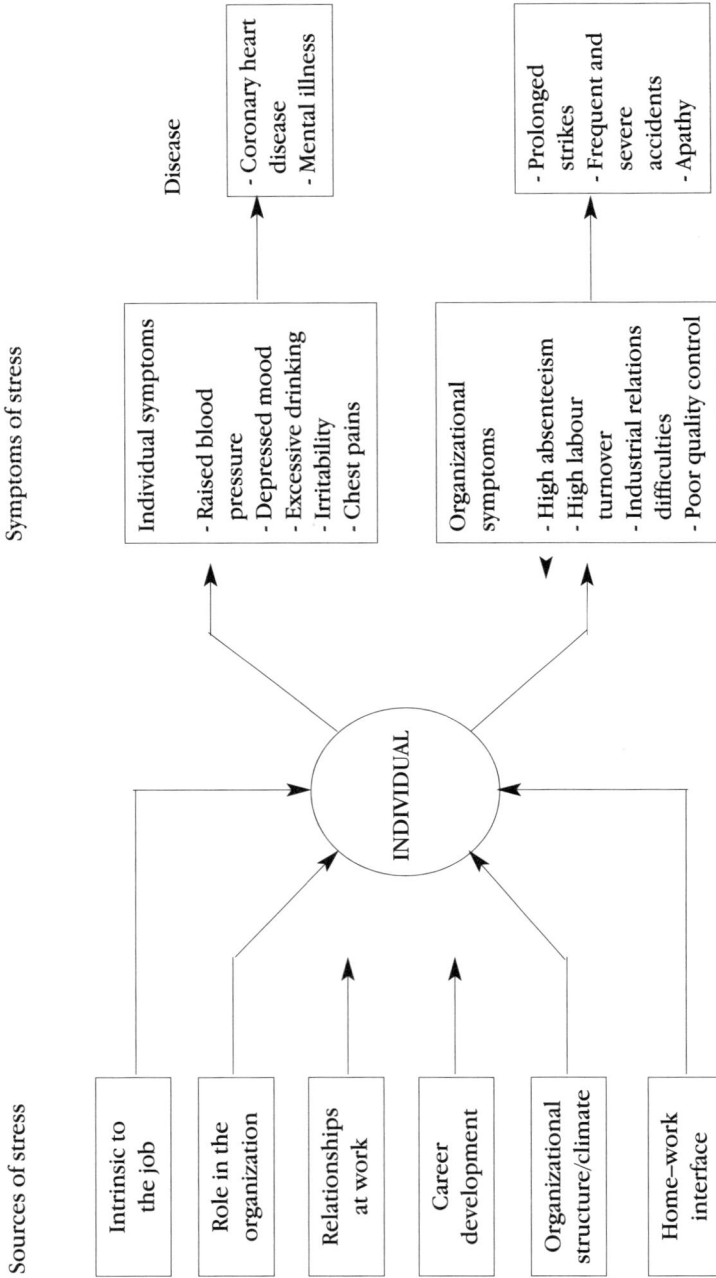

Figure 5.2: Model of occupational stress.

harm or threat to the individual, stress will result. However, if they are positively appraised and change is viewed as presenting enhanced job opportunities, then stress will be minimal. Alternatively, there may be situations when the M&A event will be appraised as being irrelevant to the individual. The uncertainty, duration and imminence of the events involved will influence the intensity of the appraisal. Cartwright and Cooper (1996) assert negative appraisals tend to dominate due to a 'fear the worst' syndrome developed from collective anxiety.

Marks and Mirvis (1986) refer to the universality of the negative response to M&A as the 'Merger Syndrome', which is characterized by heightened self-interest. Mergers are identified as being particularly stressful because they result in increased workload, uncertainty about the future and job insecurity. Marks and Mirvis (1986) draw on general stress research to emphasize that stress is determined by perceptions and not reality itself. Additionally, although stress increases vigilance in gathering information, it also leads to simplification and distortion of what employees hear. This refers to the effect of merger 'horror stories' and rumours. They argue that a crisis management mentality centralizes M&A decision-making, which isolates senior decision-makers from the workforce and so gives rise to rumours. The 'Merger Syndrome' was initially presented in 1986, and a 1997 update on the issues reveals some specific cultural/contextual variables that augment individuals' stress and their ability to cope with the change (Marks and Mirvis, 1997). These include the increase in general cynicism/distrust of leadership, 'change wariness' due to increasing change in the workplace and doubts regarding its effectiveness, and burnout from consistency of general work stress due to the rapid evolution of organizational life.

Stress theories and models significantly contribute to our under-standing of how employees respond to major change events. They allow a number of factors to be acknowledged, integrating situational elements with dispositional elements to explain individual response, and consider the central cognitive appraisal process in determining this. However, the sequential aspect of the stage models is largely neglected with the stress framework, although Ivancevich et al. (1987) do outline the different phases in the merger process and the different stressors associated with each of these. The stress models have also suffered criticism for their strong focus on negative outcomes and their failure to account adequately for positive experiences of the change process (Nicholson, 1990).

In a merger situation it is generally hypothesized that high stress levels will be manifested in behavioural outcomes such as withdrawal, absence, lateness, turnover, compensation claims and reduced productivity which will negatively affect merger performance (Hogan and Overmyer-Day,

1994; Mann, 1996). Existing research supports the assertion of elevated stress levels in M&A. For example, Siu, Cooper and Donald (1997) studied employees from an acquired Hong Kong television company and found significantly high stress levels. This was attributed to the impact of the acquisition, although this was speculative rather than empirically determined, as the study was cross-sectional. However, in a longitudinal study of an organizational consolidation, Begley (1998) reported an increase in mental distress post-consolidation from pre-consolidation levels. This can be ascribed to the changes with some confidence as pre and post measures were employed. Utilizing a control group, Gibbons (1998) compared lecturers' stress levels in colleges that had been reorganized and those that had not and found those that had undergone change reported significantly higher stress levels. Significantly high stress levels were also reported in a post-merger study of building society managers (Cartwright and Cooper, 1996). The study found that an abnormally high percentage of managers scored higher than psycho-neurotic outpatients on the CCEI, a clinical measure of mental health.

Merger-related sources of stress

A major source of stress in any change situation is the uncertainty, demonstrated as often being more stressful than the change itself (Schweiger and DeNisi, 1991; Cartwright and Cooper, 1996). Ashford (1988) studied employees coping with their company's recent divestiture and noted that perceived uncertainty and fears about the impact of the transition were related to employee stress. Uncertainty has also been documented as a pre-merger concern (Burlew, Pederson and Bradley, 1994). Employees often cope with uncertainty by reducing commitment levels and utilize this energy to cope with the anxiety and confusion (Fulmer and Gilkey, 1988). This attitude can spread and become endemic in employees, which supports Schweiger and DeNisi's (1991) conclusion that the negative effects of mergers seem to become more serious, rather than fade away, over time.

There are other more specific sources of merger-related stress. The six major sources will now be discussed within the context of the Cooper-Marshall model (1978).

Factors intrinsic to the job

M&As invariably result in increased workload, particularly when there is a significant amount of rationalization to achieve the projected economies of scale (Marks and Mirvis, 1986). Existing skills and expertise may also become redundant or devalued as a result of changes in work organization and practices to the extent that the individual job holder feels inadequate or unable to perform effectively in the new situation (Cartwright and Cooper, 1996).

M&As violate and change the relationship between employer and employee in terms of the psychological contract between the two parties. Rousseau (1995) suggests that violations reduce the commitment of individuals to the organization and that the contract between employer and employee reverts to an essentially transactional contract which focuses on equity and financial exchange. It has frequently been observed that hygiene issues tend to dominate employee concerns in the initial stages of the M&A process and not surprisingly, pay and benefits have been identified as a significant merger stressor (Covin, Sightler, Kolenleo and Tudor, 1996).

Role in the organization

Ambiguity and duplication in job roles is a consistent feature and potential source of M&A stress. Schweiger, Ivancevich and Power (1987) have highlighted loss of identity as a major post-merger concern. This sense of loss occurs when strong attachments to one's job, co-workers, routines, personal skills and performance/career goals are destroyed or changed. Even if an individual does not change jobs, there are changed role expectations through shifts in context in which individual roles are enacted (Bartunek and Fanzak, 1988). However, Schweiger et al. (1987) found that many employees accepted that redundancies, relocation and role changes were necessary, and the stressful aspect was in fact the apparently arbitrary nature of the decisions and the lack of positive regard with which these were implemented.

Work relationships

Work relationships can be an M&A stressor due to adjusting to new bosses, colleagues and subordinates. Newman and Krzystofiak (1993) observed work relationships to be affected by merger, which led to decreased job satisfaction and commitment to the organization. Siu et al.'s (1997) study of an acquired company reported relationships with others to be a strong predictor of job satisfaction, and Cartwright and Cooper (1996) assert managerial relations and personality clashes are recognized sources of merger problems.

Career development

Fear of job loss, demotion, disruption of career path, possibility of job transfer or location and loss of or reduced power, status, prestige are all common merger stressors (Cartwright and Cooper, 1996). Elliott and Maples (1991) studied individuals attending stress management workshops during an acquisition, and confirmed that worries about the future and job security were prevalent.

Concerned about their career prospects, individuals may feel on trial in an M&A situation. They may therefore attempt to promote a desirable image of themselves by working late or taking on extra responsibilities, which can exacerbate existing stress levels (Cartwright and Cooper, 1996). Similarly, Marks (1988) asserts a major cause of merger stress is fear of being perceived as vulnerable to the changed situation, and that not to be considered 'merger fit' could lead to dismissal. The stress and pretence caused by this fear can lead to long-term problematic consequences for the individual, both physically and psychologically.

Organizational structure and climate

Stressors associated with organizational structure and climate centre around poor or inconsistent communication and adjustment to changes in organizational culture. Rumours can be a powerful stressor, particularly in the absence of any formal communication from the organization. Buono and Bowditch (1989) express that rumour mills and grapevines work overtime in an M&A situation, and because rumours are often based on fears rather than reality, they can significantly exacerbate employee anxiety, tension and stress. Marks and Mirvis (1992) illustrate the exaggerated nature of merger rumours. They report that at the headquarters of an acquired optical products company a rumour of 3000 redundancies spread even though only 1700 people were employed at the site.

Other stressors associated with organizational structure and climate include lack of effective leadership. Schweiger et al. (1987) note that effective management during periods of organizational change is vital for reducing stress levels. If leaders appear to have a clear vision of the changed organization's destiny, employees are more likely to achieve a sense of control over the event (Yukl, 1989). Covin et al.'s (1996) reported satisfaction with the merger was related to agreement with the acquiring company's mission statement. It appears that not only is it significant that the company has a clear direction, but also that this is congruent with the individual's perspective. Also, the extent to which leaders empower others has been demonstrated as important in a change situation. Laschinger, Wong, McMahon and Kaufmann (1999) studied nurses following a merger of two hospitals and observed that leader-empowering behaviours significantly influenced perceptions of access to empowerment structures (information, support, resources and opportunity). Higher perceived access to empowerment structures in turn predicted lower levels of work tension for the nurses.

Home/work interface

A major change at work could have family repercussions, and obsession with this issue has been reported as a stressor (e.g. Schweiger et al., 1987). Work-related anxieties are likely to invade family life and affect relationships and the atmosphere at home. As mentioned, the need to appear 'merger fit' and survive can lead to working of longer hours and therefore adversely affect the quality of home life (Cartwright and Cooper, 1996).

Other stressors

Other stressors documented in the literature include degree of cultural change (Cartwright and Cooper, 1996), delayed change (Schweiger et al., 1987), mourning the loss of laid off co-workers, anger at their treatment and suffering from survivors' guilt for continuing to work for an organization that has treated colleagues harshly (Schweiger et al., 1987; Marks and Mirvis, 1992).

Stressors over time

Stressors are likely to differ according to the amount of time that has passed since the merger announcement. Schweiger et al. (1987) documented five short-term concerns experienced by employees following an M&A announcement: loss of identity, lack of information and increased anxiety, obsession with survival, lost talent as individuals leave the organization, and family repercussions. Marks and Mirvis (1992) assert the first reactions to merger focus on fears about jobs and careers. In the post-merger period employees will no longer be worried about future employment as their jobs will have been secured. The emphasis then alters to the scope of new tasks and responsibilities, the personalities of new bosses and co-workers and where employees stand in the new organization. They must contend with the pressures of undertaking integration tasks and simultaneously manufacturing products, serving customers, processing information and keeping the business going. Ivancevich et al. (1987) delineate the stages of the merger process more specifically. In the planning stage the possibility of job loss is a key stressor for everyone, and the next 'in play' pre-merger stage involves increasing uncertainty. Uncertainty continues during the standstill/transition period but with a shift to more specific changes, as other uncertainties are resolved. Finally, in the stabilization period the transition is complete and unforeseen changes which were not considered in the transition stage emerge. This was demonstrated in Greenwood, Hinings and Brown's (1994) research, which ascertained that unclear agreements made at early stages resulted in a cycle of escalating conflict, and ambiguities were then clarified at a later stage.

Individual and organizational outcomes

Health

There have been relatively few empirical studies that have assessed the impact of merger stress on health. As has been already mentioned, Cartwright and Cooper (1992) found evidence of poor mental health amongst managers six months post merger, even though it was a friendly merger between two culturally very similar organizations. The results also indicated that managers of the smaller merger partner were significantly more affected. Similarly, Siu et al. (1997) found that levels of job satisfaction, physical and psychological health were significantly poorer than normative data amongst acquired employees, particularly those in middle management positions. Fried, Tiegs, Naughton and Ashforth (1996) also focused on the impact of corporate acquisition on middle managers. They found that levels of experienced stress were linked to managerial perceptions of the fairness of treatment of terminated employees. In a study of a large manufacturing organization, the incidence of high blood pressure doubled from 11% in the year prior to acquisition to 22% in the first year post acquisition (Marks and Mirvis, 1997). As well as increased blood pressure, Schweiger and Ivancevich (1985) also document an increased reporting of a range of physical symptoms such as ulcers, migraines, headaches and insomnia amongst merged employees.

Attitudes

A consistent theme in the M&A literature has been the notion that it is not necessarily change in itself that causes a decline in work attitudes, but the perception of how the change is handled and communicated by the organization. Mottola, Bachman, Gaertner and Dovidio (1997) reported organizational commitment to the merged organization to be most favourable when the organization resembled neither of the pre-merger companies. Although this was a simulated study, it highlights the importance of building a new identity for the merged organization to maintain or regain employee commitment. Newman and Krzystofiak (1993) emphasize that although pre/post-acquisition measures showed a decline in job satisfaction and organizational commitment, not all employees uniformly experience decline and that it is important to identify the factors differentiating those who respond positively and negatively.

However, in summary, research suggests that job satisfaction and organizational commitment are likely to deteriorate in a change situation (Nelson et al., 1995).

Behaviours

Walsh (1988) compared the rate of staff turnover between 55 acquired organizations and 30 matched non-acquired organizations and found it to be significantly higher amongst acquired employees. Schweiger et al. (1987) report that 58% of managers in an acquired organization leave within five years of acquisition. Similarly, Baytos (1986) estimated that 50–75% of key managers from acquired organizations voluntarily leave the new organization within the first three years. Based on evidence of several M&A case studies, Cartwright and Cooper (1996) report abnormally high rates of both turnover and absenteeism amongst blue-collar and shop floor workers as well as managers.

Hambrick and Cannella (1993) explain high turnover rates amongst acquired executives in terms of 'relative standing', which refers to the extent to which acquiring managers behave in a dominant manner towards their newly acquired colleagues. Acquired executives may experience diminished relative standing if the acquiring managers behave in a dominant manner. This can have behavioural consequences, such as turnover. Hambrick and Cannella (1993) obtained mixed support for the construct. However, it was supported by Lubatkin, Schweiger and Weber (1999) in a pre/post merger study which found a causal relationship between 'relative standing' and quitting behaviour. The researchers also acknowledge that there are many other influences on turnover, such as how near acquired managers are to retirement age.

As in the case of occupational stress research generally, the direct relationship between merger stress and actual workplace performance has been little investigated and it still remains an area of hypothetical extrapolation and anecdotal reporting (Altendorf, 1986).

Acquired and acquiring employees

Employees may face different stressors according to whether they belong to the acquiring or acquired organization. Even in a merger situation there is invariably a dominant merger partner, and research suggests employees from the acquired or smaller merger partner may be confronted with more stress, possibly because they face more extensive and more negative change (Hogan and Overmyer-Day, 1994). Hambrick and Cannella (1993) discovered acquired employees perceived themselves as 'sold' as commodities and felt worthless and inferior due to loss of autonomy and status. Covin et al.'s (1996) research also signifies that acquired employees may face more stressors. They studied 2845 employees from a large manufacturing company and documented significant differences between acquiring firm and acquired firm employees in satisfaction with the merger. Employees of the acquired company experienced high levels of dissatisfaction with the merger, which implies they experienced more stress due to the changes. Covin et al. (1996) consider that antagonism and hostility may be exacerbated if the acquired organization has competed directly with the acquirer. This hostility and conflict may persist, as observed by Napier, Simmons and Stratton (1989), who studied the merger of two banks and found that even after 10 months the employees still spoke of 'we' and 'they'.

From a cultural perspective, Buono and Bowditch's (1985) study of a banking merger found that the former employees of the displaced culture were less satisfied and committed than those of the retained culture, despite holding more favourable pre-merger attitudes. Given that the dominant partner or acquirer's culture is more likely to be adopted by the new organization, this also corroborates the existence of inter-group differences. Ivancevich et al. (1987) suggest that commitment amongst the acquired workforce drops when their culture is forced to change dramatically. Terry and Callan (1998) observed group differences pre-merger. They studied high and low status hospitals that were intending to merge and revealed clear evidence of in-group bias, particularly among lower status hospital employees.

Contrary to the research discussed, Terry, Callan and Sartori (1996) found that employees of the acquired company had the most positive reactions to the merger of two airlines. This supports predictions from social identity theory, which attributes the positive reactions to the opportunity the merger presents for improvement of social identity. Dominant group employees felt their social identity as members of a prestigious international airline was undermined by inclusion of domestic airline employees in the newly formed company. These results highlight that an understanding of employees' reactions to M&A is also contingent upon pre-merger circumstances.

Occupational level

Responses to a merger may be affected by occupational level and some research has addressed this. There may be particular issues of greater pertinence to certain groups than others, and the merger politics may differentially influence staff. Nelson et al. (1995) related that the decline in job satisfaction, mental health and physical health was greatest for manual workers, which could be explained by their lower levels of control. However, managerial staff may experience different stressors. Manghan (1973) noted that managers who objected to post-acquisition change were labelled 'resistant to change' and later replaced, so this may be a particular M&A stressor for managerial staff. Siu et al. (1997) recorded a strong association between managerial role and mental ill health. However, Terry and Callan (1998) documented no significant differences in levels of threat between managers and supervisors, contrary to expectation.

Cartwright and Cooper (1996) emphasize the vulnerability of senior managers, particularly in acquired organizations. Holbeche (1998) suggests this can affect those they manage, in that because senior managers feel threatened, they focus on themselves before considering other employees. Once their jobs are secure, they have difficulty under-standing others' negative reactions because they themselves only perceive the opportunities presented by the merger. This reasoning may explain Hunsaker and Coombs' (1988) finding in their 'Merger Emotions Syndrome' research, that most executives were optimists and most staff pessimists, although this denotes personality rather than situational differences. The discrepancy in emotional styles placed the two groups in different stages of the syndrome (Figure 5.1) and may account for the perceptions that staff had that their managers lacked concern for their well-being. This draws attention to the requirement for managers to sympathize with the negative feelings of staff even if they themselves feel positive about the situation.

Fried et al. (1996) identify middle managers as a group particularly affected by an acquisition, as they have little influence on the event, but have reached a level where it is more difficult to find comparable jobs than lower level staff, or senior managers who may receive 'golden parachutes'. This research identified changes in job control and survivors' reactions to lay-offs as specific potential stressors. However, as not all middle managers were affected in the same way and to the same degree, the model proposed by Fried et al. (1996) incorporates locus of control as an impor-tant moderating variable. Additional mediating variables are helplessness and expectations of impact on career development. Within the model proposed, the variable, 'reactions to terminations' interacts with 'identifi-cation with terminated employees'; so if the individual had close relation-ships with those made redundant, then they would have a more negative reaction than if they did not. Within Fried et al.'s (1996) model, all these variables combine to account for the degree of psychological withdrawal experienced, which then translates into intention to leave.

Greenwood (1996) has praised this model for furthering insight into individual responses to change by selecting a certain group and specifying in detail the relevant factors and precisely how they interact. However, the model may not be relevant to other M&As and types of organizations, as research suggests significant differences can exist between organizations (Greenwood et al., 1994). Nevertheless, the model greatly contributes to understanding of individual responses to M&A by providing a detailed example of relevant factors and the causal logic involved in determining their responses, for a particular group of employees.

Individual differences

The occupational stress literature has identified a range of individual factors which moderate the stress response and which include such factors as type A behaviour, tolerance of ambiguity, locus of control and coping resources. In the context of merger stress, other factors like length of service, personal financial circumstances, availability of alternative job opportunities and previous merger experience (Cartwright and Hudson, 2000) may also play an important role. However, research by Ashford (1988) implies that organizational transitions create universal stress which is little moderated by personality factors and that active attempts to cope with the situation by seeking information and feedback either failed to affect or, in some cases, actually increased stress levels. These findings imply that change-related stress is pervasive and difficult to cope with.

Few studies have examined the role and effectiveness of different coping responses to reduce the stress associated with mergers. The evidence that exists is confusing and somewhat counterintuitive. Emotion-focused coping (EFC) strategies are generally suggested to be more appropriate than problem-focused coping (PFC) strategies in circumstances when the individual cannot control the environment (Forsythe and Compas, 1987). In the context of organizational change this has been upheld by research evidence (Terry and Callan, 1998). However, in a study conducted by the same researchers (Terry, Callan and Sartori, 1996) of a recently merged airline company, EFC was found to be associated with lower job satisfaction and poorer well-being than PFC. In a subsequent study of a merger in the insurance sector, Cartwright and Hudson (2000) found that PFC was positively related to psychological and physical health but that strategies such as seeking social support and information, which are generally regarded as adaptive, were associated with poor health outcomes. Interestingly, the study found that prior merger experience played a major role in determining health outcomes. Comparative to 'first timers' those with prior merger experience did not perceive the stressful potential of the merger any differently but they used more adaptive coping strategies and perceived themselves to have greater personal influence. In turn, they had significantly better mental and physical health.

Social support

The role of social support in the context of organizational change is far from clear. Cartwright and Cooper (1996) found that talking with spouse or partner was the main strategy individuals used to cope with merger stress. They comment that whilst partner support may be useful to the affected individual, because merger-related change may have family repercussions, talking about the situation may accentuate partner anxiety and that talking to superiors may be more constructive.

Callan (1993) also suggests managerial support will be beneficial during organizational change. Supportive of this contention, research has typically discovered that work sources of support are of more value than non-work sources (e.g. Terry et al., 1996). This could be because they are near to the source of stress and therefore more able to offer relevant support than sources outside the organization.

Terry et al. (1996) reported both direct and mediated effects of supervisor support in their study of an airline merger. They suggest good relationships with supervisors are important for adjustment because supervisors can provide both instrumental and emotional support, accurately communicate to employees and therefore influence threat perceptions. This study also reported that high levels of colleague support negatively influenced adjustment. Those high in colleague support rated the event as more stressful and employed higher levels of escapism than those who perceived lower levels of colleague support. This could be because under conditions of uncertainty, discussing the change with colleagues may increase anxiety, heighten threat appraisal and encourage the use of maladaptive coping responses (Kaufmann and Beehr, 1986). Moyle (1998) also documented the beneficial effects of managerial support during organizational change. She reported that managerial support directly influenced job satisfaction and mental health during organizational restructuring.

However, other research has not detected the positive effects of social support in organizational change situations. Dekker and Schaufeli (1995) studied job insecurity during radical organizational change and concluded that neither support from colleagues, management or unions protected employees from the negative effects of this common M&A stressor.

Locus of control

Spector (1994) defines locus of control (LOC) as, 'a personality variable that concerns people's generalized expectancies that they can or cannot control reinforcements in their lives'. People who believe they can are 'internals', and those who believe outside forces control reinforcements are 'externals'. Generally it is reported that 'internals' experience greater job satisfaction than 'externals' regardless of the nature of work or organization (e.g. Achamamba and Kumar, 1989). Spector (1994) conducted a meta-analysis and reported correlations between locus of control and job strains (job dissatisfaction, stress symptoms and emotional distress). In their study of Chinese employees, Siu et al. (1997) discovered strong direct effects of locus of control, on satisfaction and intention to leave. Lu, Tseng and Cooper (1999) also found that internal LOC was related to higher job satisfaction and positive mental health in a study of Taiwanese managers. Furthermore, she found that LOC was a significant moderator of the stress–strain relationship.

The more positive outcomes for 'internals' can be explained by the fact that 'externals' are less likely to cope through active problem solving and more likely to utilize emotion-focused strategies (e.g. Terry, 1991). Nelson and Cohen (1983) criticize the locus of control construct, as they obtained low correlations between an individual's locus of control classification and their appraised control over stressful events. They stated that perceptions of control were more a function of the characteristics of the situation than of individual disposition. Additionally, in their study neither locus of control nor control perceptions moderated the effects of stressful life events on psychological disorder.

Rotter (1975) contends that generalized control expectancies have their greatest influence when the situation is ambiguous or novel, such as that of organizational change (Callan, 1993). Schweiger and Ivancevich (1985) agree, emphasizing that the absence of control and uncertainty are often major stressors in merger situations. Therefore, locus of control is of probable significance. Callan, Terry and Schweitzer (1994) noted that 'internals' had lower anxiety levels in their study of lawyers coping with organizational change. In a public sector integration (Terry and Callan, 1998) it was found that 'internals' were more likely to use problem-focused coping strategies, which could account for the beneficial outcomes. Nelson et al. (1995) studied employee reactions during a water company privatization and reported that 'externals' experienced lower job satisfaction levels when the reorganization was occurring. Reinforcing the importance of control in a change environment, Ashford (1988) reported a buffering effect of personal control, and Judge, Thoresen, Pucik and Welbourne (1999) reported a positive relationship between internal locus of control and coping with change.

In Fried et al.'s (1996) structural model of middle managers' reactions to acquisition, outlined earlier, locus of control was an important determinant of outcome by influencing perceptions of stressors. However, disputing the research revealing beneficial outcomes for 'internals', Schweiger and Ivancevich (1985) argue 'internals' are more likely to respond poorly to merger, terming them 'hot merger reactors'. This is because individuals who perceive themselves as being able to control events may find uncontrollable situations extremely stressful.

Positive and negative affectivity

Positive affectivity (PA) and negative affectivity (NA) have been described as the key dispositional determinants of affective reactions at work (George, 1992) and can be defined as, 'indices of positive and negative emotionality'. The traits approximate the dominant personality dimensions of extraversion and neuroticism respectively (Watson and Clark, 1984) and are distinct, independent and stable constructs, differentially related to other constructs.

High NA individuals tend to focus on negative aspects of themselves, other people and the world in general and are more likely to experience significant levels of distress than low NA individuals (Watson and Clark, 1984). High PA individuals have a generalized sense of well-being, consider themselves to be pleasurably and effectively engaged in both interpersonal and achievement contexts, feel self-efficacious and are more likely to experience positive emotions and moods than low PA individuals (Burke, Brief and George, 1993).

In the general stress literature, the pervasive finding is of an association between NA and various self-reported stressors and outcomes (Burke et al., 1993; Spector and O'Connell, 1994). NA has popularly been depicted as a nuisance variable in stress research. Brief, Burke, George, Robinson and Webster (1988) predicted NA would be related to self-reported job stress and job strain measures and predicted that relationships between these stressor and strain measures would be inflated considerably by NA. Results from 497 managers and professionals largely supported these predictions. Burke et al.'s (1993) results corroborated the hypothesized 'nuisance' properties of NA, as NA influenced the magnitude of the correlations between the self-reports of stressors and strains. They recommend that NA should be controlled for before correlating measures of stressors and strain.

However, contradictory evidence exists. Chen and Spector (1991) demonstrated that when NA was statistically controlled, there was no significant shift in the stressor-strain relationship. Jex and Spector (1996) and Hart, Wearing and Headey (1995) reported similar results. Fogarty, Machin, Albion, Sutherland, Lalor and Revitt (1999) revealed that when NA was statistically controlled the stress–strain relationship did not disappear, it was simply reduced. Chen and Spector (1991) attest that inconsistent results could be attributed to different measures used. Fogarty et al. (1999) offer additional explanations in that some studies focus on chronic stressors while others concentrate on acute stressors, and that there are methodological shortcomings with the partial correlation approach widely used for controlling NA.

Few M&A or organizational change studies have examined affectivity as a dispositional factor influencing coping with change. In their privatization study Nelson et al. (1995) noted that neuroticism significantly contributed to the prediction of mental and psychosomatic health, suggesting individuals who are more neurotic are more vulnerable to the stressors of organizational change. Given the comparability of neuroticism to NA, this finding implies NA should be considered when explaining responses to change. Also, the optimistic and pessimistic individual differences presented by Hunsaker and Coombs (1988) in conjunction with their 'Merger Emotions Syndrome' (see Figure 5.1), could be equated with PA and NA.

Terry et al. (1996) studied employees of a newly merged airline and developed a model explaining employee adjustment to organizational change. They postulate the determinants of adjustment are work characteristics (consultation, communication, leadership), coping resources (negative affectivity, supervisor support, colleague support), situational appraisal (stress, situational control, self-efficacy) and coping responses (problem and emotion focused). This was tested among 662 fleet staff 3 months after implementation of the major changes and the results validated the model. Specifically regarding NA, which was measured with a neuroticism index, direct effects were observed for psychological well-being. In addition, a significant mediating mechanism through situational appraisals was identified for psychological well-being and job satisfaction. This study was cross-sectional so no definite conclusions can be drawn. Nonetheless, the findings corroborate Nelson et al. (1995) and accentuate the role of NA in coping with change. Terry et al.'s (1996) study appears to be the only merger study to examine NA in the existing literature.

Summary

Clearly there is a need for more research studies in the field of M&As to improve the representativeness and generalizability of current findings. Hogan and Overmyer-Day (1994) contend that future research should attempt to translate existing stress and anxiety measures into the M&A situation and adopt longitudinal research designs. Fried et al. (1996) advise that research should differentiate between hostile and friendly M&A situations and examine the impact of the time period between announcement and implementation. General stress models have been useful in providing an initial framework for the study of M&A stress but this now needs to be broadened and elaborated to account for a wider range of merger-specific variables, particularly those that may moderate the stress–strain relationship.

There appears to be emerging consensus as to the potential sources of merger stress. However, in terms of tackling the problem of merger stress, primary level interventions to reduce or alleviate merger stressors are only likely to occur by increasing organizational awareness. For this to happen, research needs to demonstrate a more direct and quantifiable link between merger stress and organizational performance. On the basis of current evidence, it would seem that individual factors, such as coping, may interact in a somewhat different way in M&A situations than in other circumstances. As this has implications for the type of secondary level interventions (i.e. stress management programmes) which may be appropriate for merged or acquired employees, a greater understanding of the way in which individuals respond and cope with M&A stress is a priority issue.

References

Achamamba B, Kumar KG (1989) I-E locus of control and job satisfaction among the workers of public and private sector undertaking. Journal of the Indian Academy of Applied Psychology 15(2): 83–6.

Altendorf DM (1986) When cultures clash: A case study of the Texaco takeover of Getty Oil and the impact of acculturation on the acquired firm (Dissertation), Faculty of Graduate School, Pasadena, CA: University of Southern California, August.

Ashford SJ (1988) Individual strategies for coping with stress during organisational transition. Journal of Applied Behavioural Science 29 19–36.

Balzer S (1997) Job loss due to mergers tops list of employee concerns. Business Journal 18(3): 76.

Bartunek JM, Fanzak FJ (1988) The effects of organizational restructuring on frames of reference and co-operation. Journal of Management 14(4): 579–92.

Baytos L (1986) The human resources side of acquisitions and divestitures. Human Resource Planning, 9(4): 167–75.

Begley TM (1998) Coping strategies as predictors of employee distress and turnover after an organizational consolidation: A longitudinal analysis. Journal of Occupational and Organizational Psychology 71(4): 305.

Brief AP, Burke MJ, George JM, Robinson BS, Webster J (1988). Should negative affectivity remain an unmeasured variable in the study of job stress? Journal of Applied Psychology 73(2): 193–8.

Buono AF, Bowditch JL (1985) When cultures collide: The anatomy of a merger. Human Relations 38(5): 477–500.

Buono AF, Bowditch JL (1989) The Human Side of Mergers and Acquisitions. San Francisco, CA: Jossey-Bass.

Burke MJ, Brief AP, George JM (1993) The role of negative affectivity in understanding relations between self-reports of stressors and strains: A comment on the applied psychology literature. Journal of Applied Psychology 78(3): 402–12.

Burlew LD, Pederson JE, Bradley B (1994) The reaction of managers to the pre-acquisition stage of a corporate merger: A qualitative study. Journal of Career Development 21(1): 11–22.

Callan VJ (1993) Individual and organizational strategies for coping with organizational change. Work and Stress 7(1): 63–75.

Callan VJ, Terry DJ, Schweitzer R (1994) Coping resources, coping strategies and adjustment to organisational change. Direct or buffering effects? Work Stress 8(4): 372–83.

Cartwright S, Cooper CL (1996) Managing Mergers, Acquisitions and Strategic Alliances: Integrating People and Cultures. Oxford: Butterworth-Heinemann.

Cartwright S, Cooper CL (2000) HR Know-How in Mergers and Acquisitions. London: IPD.

Cartwright S, Hudson SL (2000) Coping with mergers and acquisitions. In: Burke R, Cooper CL (eds) The Organization in Crisis: Downsizing, Restructuring and Renewal. London: Blackwell. (in press).

Chen PY, Spector PE (1991) Negative affectivity as the underlying cause of correlations between stressors and strains. Journal of Applied Psychology 76(3): 398–407.

Clarke H, Chandler J, Barry J (1996) Work psychology, women and stress: Silence, identity and the boundaries of conventional wisdom. Gender, Work and Organization 3(2): 65–78.

Cooper CL, Marshall J (1978) Executives Under Pressure. London: Macmillan.

Covin TJ, Sightler KW, Kolenleo TA, Tudor RK (1996) An investigation of post-acquisition satisfaction with the merger. Journal of Applied Behavioral Science 32(2): 125–42.

Cummings T, Cooper CL (1979) A cybernetic framework for the study of occupational stress. Human Relations 32: 395–419.

Dekker SWA, Schaufeli WB (1995) The effects of job insecurity on psychological health and withdrawal: A longitudinal study. Australian Psychologist 30(1): 57–63.

Elliott TR, Maples S (1991) Stress management training for employees experiencing corporate acquisition. Journal of Employment Counselling 28(3): 107–14.

Fogarty GJ, Machin A, Albion MJ, Sutherland LF, Lalor GI, Revitt S (1999) Predicting occupational strain and job satisfaction: The role of stress, coping, personality and affectivity variables. Journal of Vocational Behavior 54: 429–52.

Forsythe P, Compas BE (1987) Coping with stress during childhood and adolescence. Psychological Bulletin 101: 393–403.

Fried Y, Tiegs RB, Naughton TJ, Ashforth BE (1996) Managers' reactions to a corporate acquisition: A test of an integrative model. Journal of Organizational Behavior 17: 401–27.

Fulmer RM, Gilkey R (1988) Blending corporate families: Management and organisational development in a post-merger environment. Academy of Management Executive 2(4): 275–83.

George JM (1992) The role of personality in organizational life: Issues and evidence. Journal of Management 18(2): 185–213.

Gertsen M, Søderberg A, Torpe JE (eds) (1998) The International Dimension of Mergers and Acquisitions. Berlin: De Guyter.

Gibbons C (1998) An investigation into the effects of organisational change on occupational stress in further education lecturers. Journal of Further and Higher Education 22(3): 315.

Greenwood R (1996) Managers' acquisitions to a corporate acquisition: comment on Fried, Tiegs, Naughton and Ashforth. Journal of Organizational Behavior 17: 248.

Greenwood R, Hinings CR, Brown J (1994) Merging professional service firms. Organization Science 5: 239–57.

Hall PD, Norburn D (1987) The Management Factor in acquisition performance. Leadership and Organization Development Journal 8 (3): 23–30.

Hambrick DC, Cannella AA (1993) Relative standing: A framework for understanding departures of acquired executives. Academy of Management Journal 36(4): 733–62.

Hart PM, Wearing AJ, Headey B (1995) Police stress and well-being: integrating personality, coping and daily work experiences. Journal of Occupational and Organizational Psychology 68: 133–56.

Hogan EA, Overmyer-Day L (1994) The psychology of mergers and acquisitions. In: Cooper CL, Robertson IT (eds) International Review of Industrial and Organizational Psychology, vol. 9. Chichester: John Wiley & Sons.

Holbeche L (1998) Scary Splice. People Management 15/10/1998.

Holmes TH, Rahe RH (1967) The social readjustment rating scale. Journal of Psychosomatic Research 11: 213–18.

Hunsaker PL, Coombs MW (1988) Mergers and acquisitions: Managing the emotional issues. Personnel 65: 56–63.

ISR (1999) International Survey Research: London Report on Mergers and Acquisitions.

Ivancevich JM, Schweiger DM, Power FR (1987) Strategies for managing the issues during mergers and acquisitions. Human Resource Planning 12(1): 19–35.

Jex SM, Spector PE (1996) The impact of negative affectivity on the stressor-strain relationship: A replication and extension. Work and Stress 5: 315–23.

Judge TA, Thoresen CJ, Pucik V, Welbourne TM (1999) Managerial coping with organizational change: A dispositional perspective. Journal of Applied Psychology 84(1): 107–22.

Kaufmann GM, Beehr TA (1986) Interactions between job stressors and social support: Some counterintuitive results. Journal of Applied Psychology 71(3): 522–6.

Kessler RC, Price RH, Wortman CB (1985) Social factors in pathology. Annual Review of Psychology 36: 531–72.

Kubler-Ross E (1969) On Death and Dying. New York: Macmillan.

Laschinger HKS, Wong C, McMahon L, Kaufmann C (1999) Leader behavior impact of staff nurse empowerment, job tension, and work effectiveness. Journal of Nursing Administration 29(5): 28–39.

Lu L, Tseng H-J, Cooper CL (1999) Managerial stress, job satisfaction and health in Taiwan. Stress Medicine 15(1): 53–64.

Lubatkin R, Schweiger D, Weber Y (1999) Top management turnover in related M&A's: an additional test of the theory of relative standing. Journal of Management 25(1): 55.

Makin P, Cooper C, Cox C (1996) Organisations and the Psychological Contract. London: BPS Books.

Manghan I (1973) Facilitating interorganisational dialogue in a merger situation. Journal of Interpersonal Development 4: 133-47.

Mann SE (1996) Employee stress an important cost in mergers. Business Insurance 30(48): 24.

Marks ML (1988) The Merger Syndrome: The human side of corporate combinations. Journal of Buyouts and Acquisitions, Jan/Feb, 18–23.

Marks ML, Mirvis PH (1986) The merger syndrome. Psychology Today 20(10): 36–42.
Marks ML, Mirvis PH (1992) Rebuilding after the merger: Dealing with 'survivor sickness'. Organizational Dynamics 21(2): 18–32.
Marks ML, Mirvis PH (1997) Revisiting the merger syndrome: Dealing with stress. Mergers and Acquisitions 31(6): 21.
McManus ML, Hergert ML (1988) Surviving Merger and Acquisition. Chicago, IL: Scott, Foresman & Co.
Mottola GR, Bachman BA, Gaertner SL, Dovidio JF (1997) How groups merge: The effects of merger integration patterns on anticipated commitment to the merged organization. Journal of Applied Social Psychology 27(15): 1335–58.
Napier N, Simmons G, Stratton K (1989) Communication during a merger: The experience of two banks. Human Resource Planning 12(2): 105–22.
Nelson A, Cooper CL, Jackson PR (1995) Uncertainty amidst change: The impact of privatization on employee job satisfaction and well-being. Journal of Occupational and Organizational Psychology 68: 57–71.
Newman JM, Krzystofiak FJ (1993) Changes to employee attitudes after an acquisition: A longitudinal analysis. Group and Organization Management 18(4): 390.
Nicholson N (1990) Transition cycle: Causes, outcomes, processes and forms. In: Fisher S, Cooper CL (eds) On the Move: The Psychology of Change and Transition. Chichester: John Wiley.
Rotter JB (1975) Some problems and misconceptions related to the construct of internal versus external control of reinforcement. Journal of Consulting and Clinical Psychology 43: 56–67.
Rousseau D (1995) Psychological Contracts in Organizations. Los Angeles, CA: Sage.
Schweiger DM, DeNisi AS (1991) Communication with employees following a merger: A longitudinal field experiment. Academy of Management Journal 34(1): 110–35.
Schweiger DM, Ivancevich JM (1985) Human resources: The forgotten factor in mergers and acquisitions. Personnel Administrator Nov: 47–61.
Schweiger DM, Ivancevich JM, Power FR (1987) Executive action for managing human resources before and after an acquisition. Academy of Management Executives 2: 127–38.
Siu O, Cooper CL, Donald I (1997) Occupational stress, job satisfaction, and mental health among employees of an acquired TV company in Hong Kong. Stress Medicine 13(2): 99–107.
Spector PE (1994) Using self report questionnaires in OB research: A comment on the use of a controversial method. Journal of Organizational Behavior 15(5): 385–92.
Spector PE, O'Connell BJ (1994) The contribution of personality traits, negative affectivity, locus of control and Type A to the subsequent reports of job stressors and job strains. Journal of Occupational and Organizational Psychology 67: 1–12.
Terry D (1991) Coping resources and situational appraisals as predictors of coping. Personality and Individual Differences 12: 1031–47.
Terry DJ, Callan VJ (1998) In-group bias in response to an organizational merger. Group Dynamics 2(2): 67–81.
Terry DJ, Callan VJ, Sartori G (1996) Employee adjustment to an organizational merger: Stress, coping and intergroup differences. Stress Medicine 12(2): 105–22.
Walsh JP (1988) Top management turnover following mergers and acquisitions. Strategic Management Journal 9: 863–7.

Walsh J (1999) Merged steel giants aim to sidestep HR policy pitfalls. People Management, 17/6/1999, pp. 12.

Watson D, Clark LA (1984) Negative affectivity: The disposition to experience aversive emotional states. Psychological Bulletin 96: 465–90.

Yukl GA (1989) Leadership in Organisations. Englewood Cliffs, NJ: Prentice-Hall.

Chapter 6
Stress in the Financial Sector

HOWARD KAHN

Introduction

In the late 1980s the American Institute of Stress produced a list of the ten most stressful jobs in the United States, namely, inner-city high school teacher, police officer, miner, air-traffic controller, medical intern, *stockbroker*, journalist, customer service/complaint department worker, waitress, and secretary (*Newsweek*, 25 April 1988, pp. 40–5). The suggestion was made that these jobs were stressful because they carried responsibility without control. In the UK, Wilby (1985) reported a rating by 'stress experts' of various jobs on a ten-point scale. In that list, six occupations were grouped under the category of 'financial areas', viz. accountancy (4.3), banking (3.7), building societies (3.3), insurance (3.8), actuary (3.3.) and *stockbroker* (5.5), where a score of 1 = low and 10 = high stress. The National Business Employment Weekly Jobs Rated Almanac (Krantz, 1995) examined 250 jobs and concluded that the 25 most stressful jobs (in the US) are US president, fire-fighter, senior corporate executive, Indy class race car driver, taxi driver, surgeon, astronaut, police officer, NFL football player, air traffic controller, highway patrol officer, public relations executive, mayor, jockey, NCAA basketball coach, advertisement account executive, real estate agent, photojournalist, US representative/senator, *stockbroker*, fisherman, airline pilot, lumberjack, emergency medical technician, and architect.

Yet, after carrying out an extensive literature search and talking to employers and to other stress researchers, I have come to the perhaps surprising conclusion that relatively little academic research has been reported concerning the stress encountered by the stockbrokers, money market dealers, experts in financial futures and arbitrage, and the host of other specialists working in the financial sector. Perhaps few personnel

90

managers and line managers in the financial sector have thought it worth-while examining this group of employees, and employers and employees alike may have little interest in the topic and accept that stress is an inevitable aspect of the work. Or perhaps potential researchers have been unable to obtain the support they need or have been refused access to those working in the sector (who were too busy making money for themselves and their employers).

Two published studies are worth noting. First, Hunsaker and Pavett (1988) placed a one-page questionnaire in the trade magazine of the US brokerage industry. Close to 3000 replies were received. The investigators report that their results do not support the stereotype of the coke-sniffing, marijuana-puffing stockbroker. Some 48% of respondents cited alcohol as the most serious problem at their places of work. Second, in a study conducted in Norway, Rodahl (1989) compared the work of foreign exchange dealers with that of air traffic controllers. Both require the capacity to deal with several problems simultaneously and the ability to make quick decisions. The heart-rate of a foreign exchange trader was monitored through a two-hour period, which included first buying $17.4 million and selling $9.9 million, and then buying $19.7 million and selling $10.1 million. The heart rate of one trader was seen as 'surprisingly moderate, and not unlike the reactions observed in a number of air-traffic controllers'. Some degree of stress, however, was shown by the heart rate of a dealer who had bought $10 million, sold $6 million, and was awaiting the establishment of the new exchange rate for the dollar.

This lack of formal research is despite the fact that the financial markets and their employees are mentioned extensively and daily in the media, the sector is seen as a particularly 'sexy' one, and at least two major Hollywood movies relating to financial dealers have been released in recent years – 'Wall Street' (1988) and 'The Bonfire of the Vanities' (1990). Three examples of the level and type of coverage given by the media to the financial sector, and to its problems, are highlighted in the following:

1. The garden designers. The *Sunday Telegraph* of 4 July 1999 (p. 19) carried a story under the headline 'High-Fliers Quit the City Rat Race for a Stress-Free Career in Garden Design', in which it was reported that a former City banker and a stockbroker had given up their high-pressure careers and high incomes to become garden designers. It was suggested that these were examples of a rising trend among high-flying young professionals to 'downsize' themselves by swapping six-figure salaries for a less frantic and more creative life. The City banker gave up an annual salary of £250 000 to take a one-year garden design course, following which he considered that he would be fortunate to earn

more than £20 000 a year. He said he had been driven to make his move by his growing dissatisfaction with the City: 'Towards the end of my time in the City, all I did in the few hours that I saw my family was moan. I was very unhappy. I would often leave home by 6 am and often not return until 2 am the following day.' He also said that the garden design course had been full of people wanting to escape well-paid high-pressure jobs.

2. Job losses. The financial sector has seen as much activity as any in terms of company take-overs, mergers and reconstructions. Société Général and Paribas, two major French banks, announced in late March 1999 that they planned to merge, with the probable loss of 25% of their current UK workforce, amounting to about 800 to 900 of their 4000-strong staff in London. Banking and Finance employees trade union Bifu claimed that up to 6000 jobs would be lost world-wide due to this merger and pointed out that the City of London was already suffering significant job losses because of the mergers of Travelers with Citicorp, Deutsche Bank with Bankers Trust, and Union Bank of Switzerland with Swiss Banking Corporation.

3. The man who broke Barings. On 3 July 1999, Nic Leeson was released from Taneh Merah prison in Singapore. At the gates of the prison he was met by hundreds of press photographers, cameramen and reporters keen to record his first moments of freedom after three and a half years in captivity. In 1996, Leeson had been sentenced to six and a half-years for fraud. Working as a futures trader in the Singapore offices of Barings, Britain's oldest and most aristocratic bank, Leeson made unauthorized trades that lost £860 million and broke the bank. He almost brought about the collapse of the Singapore International Monetary Exchange, and in the aftermath of the scandal the Bank of England lost its regulatory powers to the newly created Financial Services Authority. Leeson was the son of a Watford plasterer and in previous years would not have been considered for employment by Barings, who were associated with Oxford and Cambridge universities, Sandhurst Royal Military Academy, and the English public schools. But during the 1980s, in the era of 'get rich quick' and 'loads of money', the British banks found that employees from the East End of London (the so-called 'barrow boys') were remarkably able to make quick financial calculations and market assessments, and predict what would happen on the foreign exchange and futures markets.

While there is little evidence of research examining stockbrokers, currency traders and the like, literature does exist about other employees working in banks, e.g. tellers and bank managers. For example, research has been reported of work carried out on stress in Japanese bank workers (Motohashi and Takano, 1995; Iwata and Suzuki, 1997), and on the stress reactions of bank directors (Bednar, Marshall and Bahouth, 1995; Seegers and vanElderen, 1996). However, there seems to be little difference between the sources of stress encountered by these groups, and the stress outcomes they report, and those of similar, customer-oriented workers and their managers. There is also some literature concerned with the potential and actual violence faced by this group of employees and its relation to stress. MillerBurke, Attridge and Fass (1999) explored how experiencing a traumatic event in the workplace affects employee physical health, mental health, personal functioning and work performance. They used a self-report methodology and mailed survey with 141 employees of 42 different bank branches that had recently been robbed, and they found that most employees had multiple negative consequences from experiencing a bank robbery while at work. Psychological, physical, work and personal areas were all affected by the robbery. Furthermore, more threatening incidents were associated with more severe consequences. Kamphuis and Emmelkamp (1998) examined the relationship between the experience of a bank robbery and its psychological consequences. Two groups of employees of a major commercial bank in the Netherlands participated in this study. One group consisted of subjects who had experienced a bank robbery and worked in high-frequency bank robbery areas; the matched control group consisted of non-robbed employees from banks in the same area. Victimized subjects displayed more signs of psychological distress than the control subjects, though distress decreased over time. A more medically oriented approach was carried out by Fischer, Wik and Fredrikson (1996), who exposed six bank officials to a video of a jointly experienced armed bank robbery and to a control video. They conclude that the stress induced by visual re-experience of a robbery is associated with altered activity in the paralimbic and cortical brain regions, which is of relevance for cognition and affect.

Stress in London currency dealers

The only major academic reports relating to the stress encountered by those working in the financial sector (in the City of London and Wall Street) appear to be those published by Howard Kahn and Cary Cooper (Kahn and Cooper, 1990, 1991a, 1991b, 1992, 1993, 1996), and with Sharon Elsey (Kahn, Cooper and Elsey, 1994). There were at least two reasons we became interested in this group of employees.

First, since the mid-1980s, two topics had regularly found their way into the media and been the subject of much discussion: occupational stress and financial dealers. We hoped to bring together these two topics – stress and money market dealers and traders. Second, we had read the reported comments of the manager of a US Dealing Room, 'We can trade $100 million in seven minutes or we can do absolutely nothing for seven hours. I let them go to the bathroom then' (Gray, 1983). What were the stress levels of the dealers who worked in such an organization? Cary Cooper and I decided to examine the stress problems faced by those working in the financial institutions of the City of London, i.e. those working in the UK clearing banks, the London International Financial Futures Exchange, and in merchant banks and foreign-owned banks. These are the people employed in buying and selling currency, Eurobonds and gilts, trading in stocks, fixed income institutional sales and Swaps, and providing economic analyses and research for their organizations. The typical customers of currency dealers and traders are foreign mutual funds, off-shore money managers, commodity pool operators, wealthy individuals, and multinational corporations. Some customers are specula-tors who trade currency like any other commodity, while others are 'hedgers' trying to protect foreign profits or lock into current currency prices for some future transaction. We planned to highlight the sources of work stress that may result in high levels of mental ill health, low levels of job satisfaction, and high alcohol intake. We wanted to look at the person-alities of dealers, and the ways in which they cope, or do not cope, with the stress of their job.

Our first challenge was to gain access to the Dealing Rooms. While many of the managers we contacted were only too willing to participate in the study, a number of managers were less supportive. An explanation for this reluctance has been put forward by Giles (1987), who points out that some personnel managers have awarded themselves between 7 and 10 points on a 10-point stress scale (where a higher score indicates higher perceived stress), and awarded their average member of staff 3 or 4 points. Personnel managers justify these ratings by arguing that lower-level staff do not have the same weight of responsibilities as themselves, and though their subordinates might *feel* more stressed, they were not so in reality.

Eventually, ten financial institutions in the City of London agreed to participate in the study, five of which were British-owned, two were major European International banks, and three were very large US-owned inter-national commercial banks. A total of 26 interviews were carried out with dealers. Sixteen of these interviews were recorded, four interviews were conducted without recording, and six interviews were carried out with dealers at their desks, as they worked. We then designed a questionnaire to obtain the data which would meet the requirements of the project.

The questionnaire consisted of a biographical section; the locus of control, mental health and coronary-prone (type A) behaviour scales of the Occupational Stress Indicator (OSI) (Cooper, Sloan and Williams, 1988); three scales of the Crown–Crisp Experiential Index (Crown and Crisp, 1979), i.e. free-floating anxiety, the somatic concomitants of anxiety, and depression; job satisfaction, as measured by the 15-item inventory designed by Warr, Cook and Wall (1979); extraversion/introversion and neuroticism/ stability were each measured by 24 questions from the Eysenck Personality Inventory (Eysenck and Eysenck, 1964). Sources of stress in the dealer's job were measured by the sources of stress inventory of the OSI, which is based upon the six categories identified by Cooper (1986), namely the stress associated with aspects intrinsic to the job, the managerial role, relationships with other people, career and achievement, organizational structure and climate, and the home/work interface, to which we added questions relating particularly to working in the Dealing Rooms. Stress coping was measured by the coping inventory of the OSI, which yields six scores each indicating a separate strategy for coping with stress, viz. social support, task strategies, logic, home and work relationships, time and involvement.

Two hundred and thirty-seven questionnaires were returned to the researchers, out of over 600 distributed (a near 40% response rate). Data from 225 usable questionnaires were available for analysis and a summary of some of the results obtained follows.

Job satisfaction

Compared with six other groups of employees, dealers report they are less satisfied than supervisors, managers and white-collar staff, and more satisfied than blue-collar staff, pilots and tax officers. Since it would appear to be more sensible to compare dealers with the first three groups named above, dealers do not seem to be particularly satisfied with their jobs. The highest levels of satisfaction reported were with 'fellow workers', 'amount of freedom in choosing their own working methods', 'amount of responsibility given' and 'amount of variety in the job'. The importance of 'fellow workers' is supported by comments made by dealers when we interviewed them, that dealing is a group or team activity, and the success or otherwise of a dealer is in part determined by the amount and quality of support from colleagues. Those responsible for the selection of Dealing Room staff also emphasized the importance of team work, and this was reflected in their recruitment and selection methods.

Staff working for US-owned organizations indicated the poorest job satisfaction of the three groups in the sample (dealers employed in UK-, US- and European-owned institutions), and the difference was statistically significant.

Mental health

Results from the Occupational Stress Indicator measure of mental health indicate that dealers' overall mental health is poorer (significantly so) than the middle managers, management consultants and board directors with whom they may be working, and is poorer (again significantly so) than blue-collar factory workers. Compared with the mean score of the 7526 people who had completed this inventory, dealers' overall mental health is about average. So we suggest that dealers are as mentally healthy as the general population. The three measures of the Crown–Crisp Experiential Index (CCEI) used in the present study indicate that both male and female dealers express greater free-floating anxiety and depression, but less somatic anxiety, than the general population.

The commonly held perception of dealers is that they are 'burnt-out' by the age of 35. Younger dealers (i.e. less than 35 years old) have *poorer* overall mental health than their older colleagues, as measured by the OSI. Younger dealers also indicate *more* free-floating anxiety and *less* somatic anxiety, both of which are in agreement with data presented for the wider population by Crown and Crisp (1979). Thus, it appears that dealers are no more likely to show poorer mental health as they age than does the general population. Those dealers who expect to wait longer for promotion (two or more years) show poorer overall mental health than their colleagues, and those dealers who spend longer hours working indicate poorer mental health than their co-workers. Male dealers who work longer hours show poorer mental health on all three of the CCEI mental health sub-scales; however, female dealers who work longer hours show better free-floating and somatic anxiety than their female colleagues. None of the differences between the mental health scores, however, is statistically significant.

When the mental health of dealers working in British-,US- and European-owned institutions was examined, those working in US-owned Dealing Rooms showed the poorest mental health on all scales. Perhaps the reason for this was given by one spot dealer interviewed: 'The Americans are a race of actors, everything's hyped up. They keep you on the boil all the time.' Another dealer working in a UK-owned bank told us: 'Dealers working for US and Japanese banks get calls in the middle of the night. I don't want that.' Overall mental health is poorest in US-owned Dealing Rooms, followed by British-owned, with the best scores found in European-owned Dealing Rooms. The same pattern (US poorest, British next, European best) is also shown in the CCEI free-floating anxiety sub-score. In the somatic anxiety and depression sub-scales, US dealers again display poorest scores, but European dealers are found in the middle range, with British dealers showing best scores. US dealers express significantly greater depression than their British and European colleagues.

Coronary-prone behaviour

Dealers indicate coronary-prone behaviour similar to middle managers, management consultants and board directors. They show higher coronary-prone behaviour scores than occupational psychologists, blue-collar factory workers and nurses dealing with the mentally handicapped, and as a group exhibit more coronary-prone traits than much of the population. One reason for this is undoubtedly that of self-selection, i.e. individuals who exhibit high coronary-prone behaviour may be attracted to and are selected by managers for the kind of jobs found in Dealing Rooms, that is, jobs high in responsibility and power, where the decisions made by the individual dealer have far-reaching effects. In addition, it is probably true that individuals low in coronary-prone behaviour do not succeed in the Dealing Room environment and either are perceived as failures by their superiors and 'let go', or decide for themselves early on in their careers that they do not wish to continue in the job. Whatever the reasons, we found that 63% of dealers scored higher on the measure (and so indicated higher coronary-prone behaviour) than the mean score of the general population.

Locus of control

Dealers score significantly higher on the locus of control scale than norma-tive data. This suggests a more 'external' locus of control. Dealers feel that they have less control over events than those in management positions. Consequently, it may be expected that dealers are more anxious and less able to deal with frustrating events: they would also be expected to be less psychologically healthy than those in management positions.

Extraversion

Comparison of dealers' extraversion scores with those of the normal population indicates that dealers are significantly more 'extravert' than average. Indeed, only three out of 22 employment groups who were examined by the originators of this test of extraversion (Eysenck and Eysenck, 1964) are found to be more extravert than dealers (i.e. salesmen, apprentices and student occupational therapists). One spot dealer told us: 'The job makes you extravert. It's not necessary for the job, but you get hyped up sometimes.'

Neuroticism

Dealers score higher on the neuroticism scale than the normative population, but the difference is not statistically significant. Dealers as a group are no more neurotic (as described by Eysenck and Eysenck, 1964) than the average person.

Sources of stress in the job

Dealers report that the item that is most stressful to them is 'misreading the market'. This is by far the source of most stress for dealers compared with the second-most stressful item, 'having far to much work to do'. The effect of misreading the market was highlighted by one dealer: 'When you've got the right position and it's going your way there's no need to panic, you sit back and let it happen. When you're caught out it's not a nice feeling, you want to work all the harder to get it back, to cut losses and make a profit.' Dealers report that they are more stressed than managers and board directors. Thus, it may be fair to suggest that dealers are more stressed than the people who manage them.

When we examined sub-groups of dealers, we found a number of significant differences. Female dealers reported that they had more stress caused by 'relationships with other people', by 'career and achievement', and by the 'organizational structure and climate', than male dealers; dealers aged over 35 years were more stressed by 'career and achievement' than their younger colleagues; non-graduates were more stressed by the 'managerial role' than graduates; those dealers who had responsibility for others in the Dealing Room were more stressed by the 'managerial role', by 'relationships with other people', and by the 'organizational structure and climate'; dealers who spent at least two hours per day travelling to and from work were more stressed by 'factors extrinsic to the job'; and dealers who worked in US-owned institutions reported that they were more stressed by factors 'intrinsic to the job' and by 'career and achievement' than their counterparts in UK-owned Dealing Rooms. Dealers in US-owned Dealing Rooms showed the most stress in four of the six major stress categories examined.

In summary, analysis of the figures suggests that female dealers, dealers with managerial-type responsibilities, and dealers working in US-owned organizations may encounter major problems with the stress of Dealing Room work.

Coping with stress

By far the two most popular methods used by dealers to cope with stress were by 'dealing with the problems immediately as they occur', and 'having stable relationships'. However, dealers made less use of all six coping strategies than did the normative population. This suggests that dealers either felt that they needed to make less use of these methods of dealing with stress, or that they were unable to make use of them.

Stress on Wall Street

In a follow-up study, Cary Cooper and I, along with our colleague Sharon Elsey, who was working on Wall Street, considered that it would be interesting to carry out a comparison of the occupational stress of dealers working on Wall Street with that of dealers working in the City of London (Kahn et al., 1994). In addition, it was hoped to determine whether our findings concerning stockbrokers working in American-owned financial institutions based in the City of London were also true of stockbrokers working for American stockbroking firms located on Wall Street. Two hundred and twenty questionnaires were distributed to Wall Street dealers, of which 57 usable questionnaires were returned, a response rate of approximately 26%.

Dealing in both New York and London is dominated by males (75% on Wall Street, 83% in the City of London), by married employees (61% and 54%, respectively), by employees aged under 33 years of age (51% and 67%, respectively), who have no children (63% in both locations), and who are relatively new in the job (51% of Wall Street dealers have been in the job for less than three years, compared with 48% of City of London dealers). Dealers on Wall Street would appear to be the better academically qualified, with 61% having a first degree (compared with 35% for the City of London), and 21% a higher degree (compared with 14%), though there may be arguments about the relative merits of UK and US academic qualifications.

Sixty-nine per cent of dealers both on Wall Street and in the City of London work up to 50 hours per week. City of London dealers state themselves to be more optimistic about their promotion prospects, with 56% expecting a promotion within one year, compared with 35% on Wall Street.

Wall Street dealers report being more stressed by all sources of stress examined, significantly so by 'relationships with other people'. They also report utilizing more of all the stress-coping techniques examined, with 'task strategies', 'time' and 'involvement' significantly so. All stress-coping techniques are used more by the Wall Street group, but only the technique of 'involvement' is used significantly more.

So what conclusions can we reach from comparing the two studies of dealers? Demographically, there are few major differences between dealers working on Wall Street and in the City of London other than those relating to their ages. Despite Wall Street dealers indicating that they perceive more job-related stress, the fact that they utilize more of the coping strategies, have a more internally oriented locus of control and a similar pattern of type A behaviour, may help them achieve better mental health than dealers in the City of London. Similar findings apply when comparing dealers on Wall Street with those in US-owned City of London Dealing Rooms. It appears that working in City of London Dealing Rooms owned by US institutions does not result in the dealers becoming like dealers on Wall Street.

These results suggest that organizations operating Dealing Rooms on Wall Street and in the City of London can expect to find similar stress sources and outcomes in both financial centres.

Comments on Kahn and Cooper's work

A number of researchers have made reference to Kahn and Cooper's examination of dealers. Their comments are included here in the hope that Dealing Room managers and researchers might find them useful.

In their large-scale nationwide study of the occupational stress found among teachers in the UK, Travers and Cooper (1993) note that the highest levels of job satisfaction expressed by teachers were with their fellow-teachers, the amount of variety in the job, the freedom to choose their own methods of working, job security and the amount of responsibility they were given. They point out that these are findings similar to those found by Kahn and Cooper (1990). Travers and Cooper go on to say that teachers made it clear that the support of their colleagues was vital when coping with stress. The same certainly does appear to be true for City dealers, and ways of encouraging this, by job design for example, should be found by their employers and managers.

An interesting use of the work carried out by Kahn and Cooper has been made by Itzhaky and Ribner (1999). They attempted to determine whether differences existed between the values of men and women who, as refugees from a fundamentalist country, were confronted by the challenge of acculturating into a Western society. They note that Kahn and Cooper (1990) cited a high level of job satisfaction among those financial dealers in the City of London having internal locus of control. From this and other studies they suggest that although some of the research they cite has dealt with migrant populations, none has considered the accultura-tion complexities of a refugee population from a fundamentalist country that had to undergo a relatively rapid adaptation to the profoundly different cultural norms of Western society. From their study, Itzhaky and Ribner (1999) conclude that the more that immigrant group members feel in control of their lives (men, in their study), the greater the potential for satisfactory experiences in the new environment. They also point out that techniques for the enhancement of internal locus of control have been developed (e.g. Aurbach, 1992). Thus, we might propose that employees in the financial sector who are required to work in other cultures might benefit from using such techniques. We came across a number of City dealers who had to undertake such a (physical) move across cultures, but whether the comparison with the group of émigrés described by Itzhaky and Ribner (1999) applies in the world's financial markets would make an interesting study.

Martin and Roman (1996) emphasize a particular aspect of Kahn and Cooper's (1990) work. They note that studies examining the relationship of job satisfaction to workers' drinking behaviours, although not numerous, do provide cursory evidence that dissatisfied workers are more prone to problems related to their use of licit and illicit substances. They give Kahn and Cooper's (1990) results as an example of this negative relationship, but go on to point out that although the negative association between job satisfaction and substance use is not strong, this may be due to the methodological limitations which have characterized the examinations carried out, for example, using small, non-random samples made up of particular occupational groups. Martin and Roman (1996) also point out that limitations relative to measurement are also seen as common in many studies utilizing single-item measures of job satisfaction and simple quantity frequency-based measures of alcohol abuse. They also comment that many studies of the effect of job satisfaction on workers' drinking behaviour have relied heavily on simple zero-order analyses, ignoring the possibility that the direct relationship between satisfaction and alcohol use may be suppressed by allowing unmeasured correlates to vary.

An examination of the longitudinal relationships between family history of alcoholism, stress levels, utilization of coping strategies, and alcohol-related problems, was carried out by Johnson and Pandina (1993). They consider from Kahn and Cooper's (1990) study that a stress coping (or a stress social resources) interaction has been demonstrated to predict more accurately subsequent levels of alcohol use and more drinking-related problems. However, Kahn and Cooper's (1990) study was not longitudinal and made no such predictions. It was a cross-sectional study and suggested that a number of dealers might run into problems in the future, but no more than this. Certainly, a longitudinal study of the drinking (and other stress-related behaviours) of employees in the financial sector would prove useful.

Mack, Nelson and Quick (1998), in examining the stress of organizational change in the global marketplace, have noted that a growing body of literature has demonstrated that occupational stress is a significant issue on an international level, and that several studies have failed to detect significant differences in stress levels between subjects in other countries and those in the US, regardless of cultural and environmental differences. They note that Kahn et al. (1994), having compared financial dealers working in the City of London and on Wall Street, found very few differences in sources of stress and the dealers' ability to deal with it. Mack et al.'s (1998) comments are important, given the growth of the global marketplace, and the increasing movement of employees working with international organizations from one national base to that located in another country. Researchers and managers particularly concerned about the need to acculturate expatriate employees should take their comments into account.

Mughal, Walsh and Wilding (1996) are critical of Kahn and Cooper's (1990) study. Reporting an investigation of the relationship between trait anxiety (TA) and stress and work performance (based on a study of 65 insurance sales consultants), they suggest that much of the theoretical framework underlying TA has evolved from laboratory research, and explicit hypotheses have rarely accompanied TA's inclusion in occupational stress research. This is certainly true of the work reported by Kahn and Cooper.

The Occupational Stress Indicator has been translated into Brazilian-Portuguese, and applied in Brazil (de Moraes, Swan and Cooper, 1993). Kahn and Cooper (1991b) are cited by them as having indicated that reliability and validity data for the English version of the OSI are of a high standard. However, while Kirkaldy, Cooper, Eysenck and Brown (1994) have noted that the Occupational Stress Indicator has shown its value as a way of identifying the major contributory factors to stress in a wide range of different occupational categories (and citing Kahn and Cooper (1991b) as an example), they also note that Kahn and Cooper (1991b) found that the OSI sub-scale of coping strategies has not been found to be uniformly reliable. Lim (1995) notes that Kahn and Cooper (1991b) did not consider the validity of the sub-scales within the coping inventory, but Kirkaldy et al. (1994) carried out a study with senior police officers to examine the psychometric properties of, inter alia, the OSI's coping strategies scale. They failed to replicate the six sub-scales of the coping scales of the OSI as proposed by Cooper et al. (1988). They suggest that a four-factor solution to the coping scale has emerged from their study, but that further empirical exploration is required.

In a review of the literature on occupational health published between the years 1991 and 1995, Kinicki, McKee and Wade (1996) have attempted to use the models of occupational health and stress they found to develop an overarching taxonomy with which to conceptually integrate the literature. They noted that Kahn and Cooper (1991a) in their study of money market dealers derived a factor labelled Technological Aspects (of a job), which predicted increased free-floating anxiety, but not somatic anxiety, depression, job satisfaction or alcohol consumption; and they also noted that Kahn and Cooper (1991a) speculated that the characteristics of their sample (highly educated, highly paid, professional employees) accounted for the insignificant results. Kinicki et al. (1996) suggest that longitudinal research with a variety of occupations using both objective and subjective measures of computer use is needed to explicate the relation between computer use and health outcomes. They also conclude that there are both methodological and theoretical issues in occupational health research, noting that of the studies they uncovered, 77% used a cross-sectional design that prohibits clear identification of the causal relations between short- and long-term outcomes, and that 68% of the studies used common measurement methods (generally self-report surveys) for both predictors and criteria. Kinicki et al. (1996) suggest that method bias may be a serious confound in occupational health research and underscore the need to use multiple measurement methods whenever possible.

In their examination of the 'politics' of international money, Thrift and Leyshon (1994) talk of the international financial system as operating in two ways. First, there is a disembedded electronic space; second, there is a re-embedded set of meeting places (from restaurants to trading floors), where many of the practices of the first space still have to be negotiated. 'The second, re-embedded space is increasingly an outcome of the first; it is an integral part of disembedded electronic space rather than a relict feature' (Thrift and Leyshon, 1994). Kahn and Cooper's (1993) examination of the modern trading 'floors' is used in support of this hypothesis. Thrift and Leyshon (1994) point out that Kahn and Cooper (1993) have stated that a number of City dealers were working class, from the east end of London (the 'barrow boys'), yet even they conformed to the dress code of the City – the suit.

Stober and Seidenstucker (1997) have developed a 'Worry Inventory for Managers'. To do this they obtained an initial item pool of the worries of managers by inspecting measures used in previous research on job stress, burnout and work environment perceptions, including Kahn and Cooper (1992). In applying their 24-item measure they found that job involvement did not show a predictive correlation with worry. In Kahn and Cooper (1990) involvement was examined in its role as a stress coping mechanism, and a suggestion was made that it was predictive of improved mental health and job satisfaction, and of less alcohol consumption. This variable is clearly worthy of further study.

Finally, in a review of Kahn and Cooper (1993), Matthews (1994) comments that the book is a 'lively and informative account of the pressures faced by financial dealers in the City of London', with the right balance between accessibility and accuracy, but that it occasionally 'fails to do justice to some of the knottier issues of the research area'. Matthews (1994) highlights two of these issues as the magnitude of significant causes, and causality. Kahn and Cooper (1993) are criticized for overlooking the point that although relationships between life events and physical and mental stress outcomes may be statistically significant, life events often predict only a small part of the variation in stress symptoms. Matthew's (1994) review comments that the results given in the book are 'limited by the cross-sectional nature of the design. Within a longitudinal design it might have been possible to look at dealers' reactions to specific events, the perspective emphasised by contemporary transactional models of stress' (pp. 573–4). (I wish others researching the stress of Dealing Room work more success than we had in gaining access. It was difficult enough to obtain agreement to carry out a cross-sectional study, let alone a longitudinal study!) Other aspects of a study of dealers which Matthews (1994) suggests should be investigated in more detail are the rewards of the job, and the reasons why dealers are prepared to tolerate the intrinsic demands of the job and organizational stressors such as role conflict. Since thrills and risks are seen as one of the attractions of the job, future work should follow this up. Although risk-taking was generally seen as a source of stress, there may be some dealers who thrive on qualities of the work which most people would find aversive, such as high uncertainty and potential for error, competition with colleagues and doubtful career paths. Such investigations are worthy of consideration.

A further criticism made by Matthews (1994) is about the use of regression analysis in the report of the study. He suggests that the authors seem to think that multiple regression provides direct access to causal relationships, which is not the case in a cross-sectional study.

'Regression may be used to test the plausibility of specific causal hypotheses *a priori*, by entering sets of variables hierarchically, in a preset order. However, the text does not indicate whether all variables were entered simultaneously, or whether a stepwise procedure was used, or whether variables were entered hierarchically ... They also describe a factor analysis ... but omit critical procedural details such as the criterion used to decide how many factors to extract, and whether or how the factors were rotated' (Matthew, 1994).

A possible reply to these latter comments is one given by Matthews (1994) himself:

'the academic researcher has a genuine problem in conveying such analyses to non-specialist readers, so I sympathise with Kahn and Cooper's reluctance to deal with the difficulties directly. However, my belief is that more effort should be made to communicate to wider audiences the limitations of multivariate techniques and the essential details of procedures used' (pp. 573–4). Kahn and Cooper (1993) is written 'for occupational health specialists, personnel managers and consultants as well as business professionals' (from the cover of the text),

and we did explain in various parts of the text the role of regression analysis. But Matthews' comments are well taken, and future researchers attempting to bring their results to a wider audience may wish to provide more details about the limitations of the methods they used. As for occupational psychologists, we would refer them to Kahn and Cooper (1990, 1991a, 1991b, 1992, 1996), and to Kahn et al. (1994).

Litigation in the financial sector

The number of cases of stress-related litigation in the UK, with employees seeking damages for illnesses caused by their employer's negligence, has increased radically in recent years. Key cases have been Walker vs. Northumberland County Council (1997) and Lancaster vs. Birmingham City Council (1999). Cartwright and Cooper (1994) in a review of the legal issues involved, concluded that the cost (to employers) of losing claims may be significant, and employees who ignore the warning signs do so at their peril. They note that 'the sources and consequences of stress in some jobs, for example ... dealers (Kahn and Cooper, 1993) are becoming increasingly documented and given publicity'.

In 1998 I was invited to act as an 'expert witness' in a case where a City of London dealer sued his ex-employer 'for damages for personal injuries occurred to the plaintiff by reason of the negligence and/or breach of statutory duty by the Defendant ... during the course of the Plaintiff's employment'.

Among the evidence put forward by the Plaintiff on his behalf was the following:

• The Plaintiff, before joining the organization he was now suing, had already suffered nervous exhaustion in a previous (similar) employment, and had been treated for 'swings of mood related to stresses particularly arising from work'. This was known to his new employer, but they had failed to take it into account in determining his workload and the demands made upon him. In order to meet the many deadlines imposed upon him, he had to work in excess of 70 hours per week, 11 hours a day, 6 days a week for long periods; he also had to work on statutory holidays. He did not have a holiday during the two and a half years preceding his psychological collapse.

- The Plaintiff was given no technical support to carry out his work, in comparison with his company's offices abroad, where assistance to those carrying out the same type of work was readily available.
- He was so successful in his job that his area of the business grew rapidly, but he was given no administrative support. A behavioural analysis of the Plaintiff, carried out by his employer, concluded that he 'should be provided with substantial administrative support'. Many senior members of staff in his organization also put it on record that he did indeed require support. The company nurse and personnel manager were both aware of his problems, and supported his requests for support. Nothing was done to help him.
- On one occasion he was required to visit the US on business and while there instructed to attend a meeting in London, which he did while suffering from jet lag and lack of sleep. There were many incidents such as this.
- He had seven changes of line manager in 30 months, each with a different management style.
- He was provided with a temporary secretary – who had no recent secretarial experience, and who left after three months; a new, very suitable secretary was supplied to him, but while the Plaintiff was away from his office on business, was 'poached 'by a senior manager.
- He made it clear to his managers that he needed time off to recover his health, but was told not to admit to suffering from 'stress' as this would prejudice, or more likely end, his career in the City.

At the time of writing, this case is still to be heard and the validity of any of the claims made has not been proven, but employers and employees in the financial sector should consider the charges made, given the potential consequences. Clearly the litigant's employers had not read Kahn and Cooper (1993), where it is written that 'the typical dealer may encounter serious health problems in the future'.

Cary Cooper's contribution to our understanding of the causes, outcomes, management and prevention of stress in general, and in the financial sector in particular, is one which should be recognized and applauded.

References

Aurbach D (1992) Locus of Control and Adaptation to an Authoritarian Organisation. Unpublished Doctoral Dissertation (in Hebrew), Bar-Ilan University, Ramat Gan, Israel.
Bednar A, Marshall C, Bahouth S (1995) Identifying the relationship between work and nonwork stress among bank managers. Psychological Reports 3: 771–7.

Cartwright J, Cooper CL (1994) Employee stress litigation: the UK experience. Work and Stress 8(4): 287–95.

Cooper CL (1986) Job distress: Recent research and the emerging role of the clinical occupational psychologist. Bulletin of the British Psychological Society 39: 325–31.

Cooper CL, Sloan SJ, Williams S (1988) Occupational Stress Indicator. Windsor: NFER-Nelson.

Crown S, Crisp AH (1979) Manual of the Crown–Crisp Experiential Index. Sevenoaks, UK: Hodder and Stoughton.

de Moraes LFR, Swan JA, Cooper CL (1993) A study of occupational stress among government white-collar workers in brazil using the occupational stress indicator. Stress Medicine 9: 91–104.

Eysenck HJ, Eysenck SBG (1964) Manual of the Eysenck Personality Index. Sevenoaks, UK: Hodder and Stoughton.

Fischer H, Wik G, Fredrikson M (1996) Functional neuroanatomy of robbery re-experience: Affective memories studied with PET. Neuroreport 7(13): 2081–6.

Giles E (1987) Stress in your own backyard. Personnel Management 19(4): 26–9.

Gray J (1983) What's behind the fastest money game of all? Medical Economics 60(24): 118–32.

Hunsaker PL, Pavett CM (1988) Drug abuse in the brokerage industry. Personnel 65: 54–8.

Itzhaky H, Ribner DS (1999) Gender, values and the work place: considerations for immigrant acculturation. International Social Work 42(2): 127–38.

Iwata N, Suzuki K (1997) Role stress-mental health relations in Japanese bank workers: A moderating effect of social support. Applied Psychology-An International Review-Psychologie AppliquÈe-Revue Internationale 46(2): 207–18.

Johnson V, Pandina RJ (1993) A longitudinal examination of the relationships among stress, coping strategies, and problems associated with alcohol use. Alcoholism: Clinical and Experimental Research 17(3): 696–702.

Kahn H, Cooper CL (1990) Mental health, job satisfaction, alcohol intake and occupational stress among dealers in financial markets. Stress Medicine 6(4): 285–98.

Kahn H, Cooper CL (1991a) The potential contribution of information technology to the mental ill-health, job dissatisfaction and alcohol intake of money market dealers – an exploratory study. International Journal of Human–Computer Interaction 3(4): 321–38.

Kahn H, Cooper CL (1991b) A note on the validity of the mental health and coping scales of the occupational stress indicator. Stress Medicine 7: 185–7.

Kahn H, Cooper CL (1992) Anxiety associated with money market dealers – sex and cultural differences. Anxiety Research 5: 21–40.

Kahn H, Cooper CL (1993) Stress in the Dealing Room – High Performers under Pressure. London: Routledge.

Kahn H, Cooper CL (1996) How foreign exchange dealers in the City of London cope with stress. International Journal of Stress Management 3(3): 137–45.

Kahn H, Cooper CL, Elsey SP (1994) Financial dealers on Wall Street and in the City of London – is the stress different? Stress Medicine 10(2): 93–100.

Kamphuis JH, Emmelkamp PMG (1998) Crime-related trauma: psychological distress in victims of bank robbery. Journal of Anxiety Disorders 12(3): 199–208.

Kinicki AJ, McKee FM, Wade J (1996) Annual Review, 1991–1995: Occupational Health. Journal of Vocational Behavior 49: 190–220.

Kirkaldy BD, Cooper CL, Eysenck M, Brown J (1994) Anxiety and coping. Personality and Individual Differences 17(5): 681–4.

Lim T (1995) Stress demands on school administrators in Singapore. Work and Stress 9(4): 491–501.

Mack DA, Nelson DL, Quick JC (1998) The stress of organizational change: a dynamic process model. Applied Psychology: An International Review 47(2): 219–32.

Martin JK, Roman PM (1996) Job satisfaction, job reward characteristics, and employees' problem drinking behaviors. Work and Occupations 23(1): 4–25.

Matthews G (1994) Book review. Journal of Management Studies 31(4): 572–5.

MillerBurke J, Attridge M, Fass PM (1999) Impact of traumatic events and organizational response – A study of bank robberies. Journal of Occupational and Environmental Medicine 41(2): 73–83.

Motohashi Y, Takano T (1995) Sleep habits and psychosomatic health complaints of bank workers in a megacity in Japan. Journal of Biosocial Science 27(4): 467–72.

Mughal S, Walsh J, Wilding J (1996) Stress and work performance: the role of trait anxiety. Personality and Individual Differences 20(6): 685–91.

Rodahl K (1989) The Physiology of Work. London: Taylor and Francis.

Seegers G, vanElderen T (1996) Examining a model of stress reactions of bank directors. European Journal of Psychological Assessment 12(3): 212–23.

Stober J, Seidenstucker B (1997) A new inventory for assessing worry in managers: correlates with job involvement and self-reliance. Personality and Individual Differences 23(6): 1085–7.

Thrift N, Leyshon A (1994) A phantom state? The de-traditionalization of money, the international financial system and international financial centres. Political Geography 13(4): 299–32.

Travers CJ, Cooper CL (1993) Mental health, job satisfaction and occupational stress among UK teachers. Work and Stress 7(3): 203–19.

Warr P, Cook J, Wall T (1979) Scales for the measurement of some work attitudes and aspects of psychological well-being. Journal of Occupational Psychology 52: 129–48.

Chapter 7
Stress and the Woman Manager

SANDRA L. FIELDEN AND MARILYN J. DAVIDSON

Over the past few decades, there have been gradual increases in the percentage of women entering the workforce in most countries in the Western hemisphere. This has also been paralleled by slight increases in the percentage of women entering various management positions (particularly in Europe, Australasia and North America), although women are still seriously under-represented at all senior executive levels (Davidson and Burke, 2000; Vinnicombe, 2000). In the UK, for example, the percentage of women managers has increased from 9.5% in 1994 to 18% in 1998. However, the majority of women are still concentrated in the lower levels of management with the percentage of women directors actually falling to 3.6% in 1998 compared to 4.5% in 1997 (Institute of Management and Remuneration Economics, 1998). This survey also highlighted the persistence of occupational segregation within management positions, with more than one-third of female managers concentrated in marketing and personnel functions. In contrast, in research and development, physical distribution, manufacturing and production, and purchasing and contracting, women only represented 6% of managers. Women still earn less at every level of management (including director level), and on average are younger than their male counterparts at each responsibility level (Institute of Management and Remuneration and Economics, 1998). Even in America, despite affirmative action legislation, the glass ceiling (a transparent barrier, keeping women from rising above certain levels in organizations) still persists at corporate executive level. In 1996, women represented only 10% of corporate officers and 1.9% of top earners (Catalyst, 1996).

A British survey by MSF Research, a section of the union for skilled and professional people, found that women still believe there is a 'glass ceiling' and that they are not afforded the same opportunities for career advancement as their male counterparts, despite organizations having equal opportunities policies (Equal Opportunities Commission, 1996). Three-quarters of women managers reported that it was easier for men to secure promotion (especially in manufacturing and computing companies), just over half blamed male 'networking' for excluding women from management positions, and 55% believed that management only paid lip service to equal opportunities (Equal Opportunities Commission, 1996). Only organizations which have a genuine commitment to equal opportunities have made any progress in removing the barriers faced by female managers, and that progress appears to be limited to a small number of organizations in traditionally female-dominated industries (Davidson and Burke, 2000). Furthermore, since the early 1980s, numerous international studies have consistently concluded that managerial and professional women experience unique and additional sources of stress related to their minority status and gender and that these stressors result in higher levels of overall occupational stress compared with their male counterparts (Cooper and Davidson, 1982; Davidson and Cooper 1983, 1992; Cooper and Melhuish, 1984; Nelson and Burke, 2000; Gardiner and Tiggemann, 1999).

This chapter, based on the model in Figure 7.1, explores the major sources of stress, both organizational and extra-organizational, encountered by women managers, and the factors that influence the responses of women managers to those stressors. It will also consider the potential impact of such stress on the behaviour and the mental and physical well-being of women managers, by evaluating the risks facing female managers as a result of their position within the workplace.

Stress

The term stress has been used to signify both environmental agents that disturb functions, as well as responses to such agents by psychological, physiological and sociological systems. It is now generally accepted that occupational stress can only be adequately explored by taking a multidisciplinary approach (Cooper, Cooper and Eaker, 1988) and while environmental factors and the reacting individual are vital elements, it is the nature of the relationship between the two that is crucial. Cartwright and Cooper (1997, p. 6) propose that 'a stress is any force that puts a psychological or physical factor beyond its range of stability, producing a strain within the individual'. Pressure in itself is not always a negative experience and can have substantial motivational benefits for those who have the resources to meet the demands placed upon them. Stress results when individuals cannot meet the demands placed upon them and as Lazarus (1971, p. 196) explains,

STRESSORS

Discrimination
Glass ceiling
Lower pay
Occupational gender segregation
Negative attitudes towards women
managers

Indirect
Exclusion from male networks
Resisting political activity

Extra-organizational
Poor domestic/emotional support
Home/family/work conflicts
Longer hours on home/family duties

Organizational culture
Male-dominated culture
Think manager – think 'white' male
Continuous work profile culture

Direct exclusion
Isolation
Token woman issues
Lack of same sex/ethnic role models
Sexism/racism
Harassment – sexual/racial

INDIVIDUAL/PERSONALITY
CHARACTERISTICS

Self-concept
Self-esteem
Self-efficacy
Control
Coping

OUTCOMES

Job dissatisfaction
Poorer job performance
Decreased career
aspirations

Physical ill health
(e.g. coronary heart
disease)

Mental ill health
(e.g. depression,
anxiety)

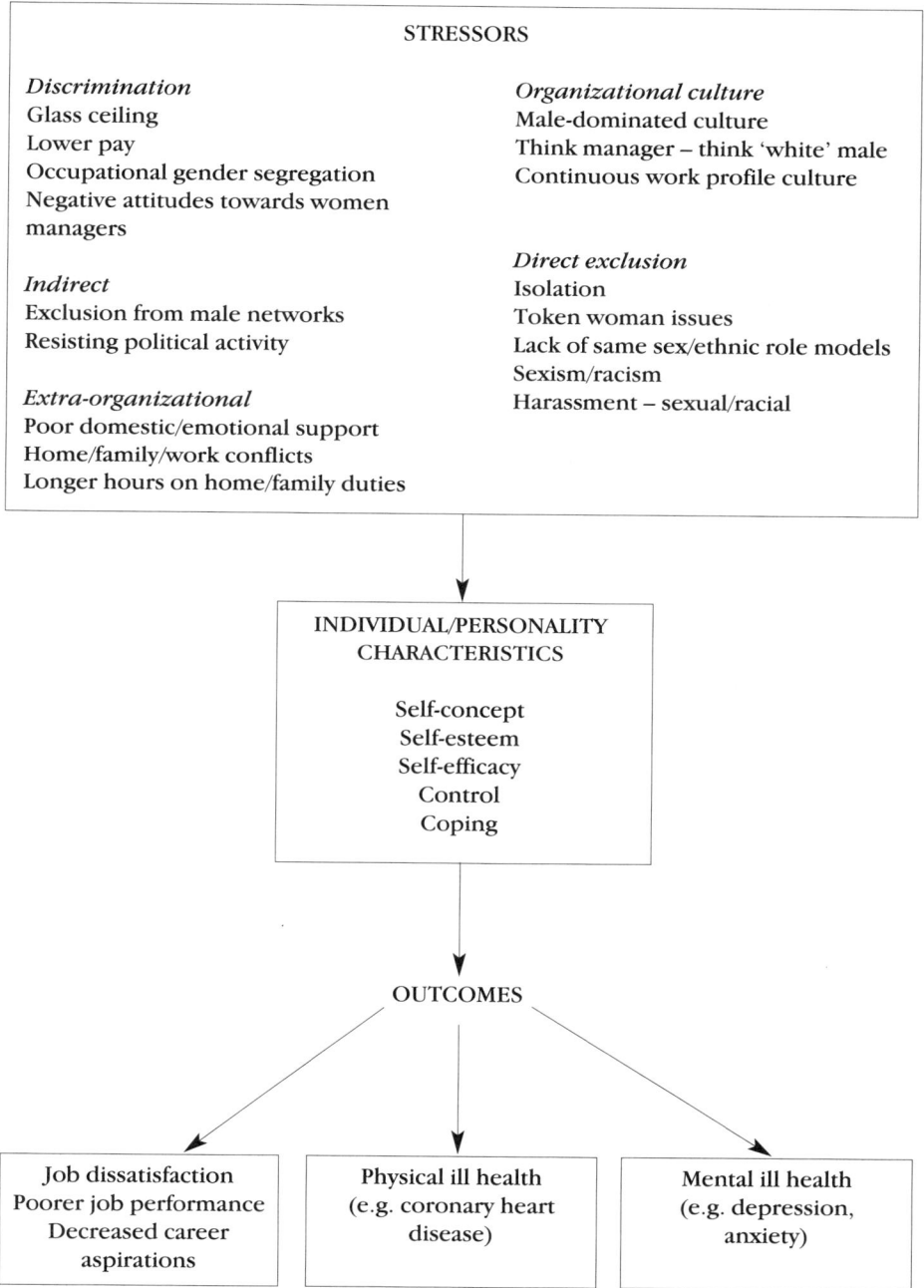

Figure 7.1: Model of Occupational Stress and the Woman Manager.

reaction depends on how the person interprets or appraises (consciously or unconsciously) the significance of a harmful, threatening or challenging event.

This definition of stress recognizes that the sources of stress, and its effects, are multiple and not just limited to a particular situation (e.g. work). It views stress as not just a function of being 'under pressure' in an occupational sense but as a function of an individual's whole life situation. Specific approaches to occupational stress also take account of an individual's psychological, physiological and behavioural responses in seeking to adapt and adjust to both internal and external pressure (e.g. Cooper and Marshall, 1976). They consider not only the sources of stress encountered in the work setting, but also the impact of external influences on an individual's reaction and perception of threat. The stress response is a highly complex one and incorporation of non-work factors into the stress equation provides a much greater understanding of this process.

Stress exerts a high price throughout the Western world. In America of the 550 million days productivity lost through sickness absence, it is estimated that 54% are stress-related in some way (Elkin and Rosch, 1990). In the United Kingdom it is estimated that the 360 million working days lost annually as a result of sickness absence are costing organizations a staggering £8 billion (O'Driscoll and Cooper, 1996). Although paid employment provides many positive benefits, the cost to women managers in terms of well-being is obviously high. Excessive pressure and scarcity of time can adversely affect women's ability to cope, often leading to reduced well-being and the increased use of maladaptive coping strategies (e.g. alcohol and drug abuse). However, a recent study by Long (1998) suggests that the impact of such stressors on female managers is mitigated by other factors. Greater access to coping resources, such as an increased sense of control, lead to higher levels of job satisfaction and lower levels of job distress for managers compared with women in clerical positions. Therefore, female managers may be less at risk than their non-managerial counterparts, but in comparison to their male managerial counterparts they are still likely to be faced with many additional sources of stress and poorer access to coping resources.

Organizational managerial stressors

Discrimination

Recent research clearly shows that while women managers do experience their jobs as challenging and stimulating, they continue to suffer additional pressures at work related to discrimination and prejudice and men are still in a more favourable position than women (Lundberg and

Frankenhaeuser, 1999). Women continue to be concentrated in the lower management levels, where they are paid less than their male counterparts yet expected to work harder (Roos and Gatta, 1999). It has been suggested that women have lower pay expectations and are easily satisfied with lower salaries, explaining why salaries in female-dominated occupations tend to be poorer than those in male-dominated occupations (Stevens, Bavetta and Gist, 1993). However, the evidence does not support the assumption that women do not value pay as strongly as men (Loscocco and Spitze, 1991). Occupational gender segregation is attributed to greater satisfaction for women because they tend to be influenced by pay comparisons with other women at similar pay levels, not because they are satisfied with less. In contrast, jobs in male-dominated occupations may afford women higher wages than those in female-dominated occupations, but they can also lead to greater pay dissatisfaction for women as they tend to be influenced by pay comparisons with male counterparts (Rosenfeld and Spenner, 1992). These pay comparisons are likely to be extremely unsatisfactory as women working in male-dominated occupations continue to receive significantly lower levels of remuneration, with salaries up to a third higher for men than for women (Reskin and Ross, 1992; Stevens et al., 1993). Those in such positions frequently report high levels of mental and physical ill health and job dissatisfaction, regardless of the type of industry or employing organization (Davidson, Cooper and Baldini, 1995).

Furthermore, the findings from a recent study carried out by Gardiner and Tiggemann (1999) suggested that both gender and the gender ratio of the industry influenced leadership style, stress and mental health in female and male managers. Women in male-dominated industries reported the highest level of pressure from discrimination. Moreover, women managers in male-dominated industries reported worse mental health when they utilized an interpersonally-oriented leadership style, whereas men managers working in male-dominated industries reported better mental health when they utilized such a leadership style. Gardiner and Tiggemann (1999) point out that these findings may help to contribute to our understanding of the barriers to women working in senior management roles in male-dominated industries, particularly keeping in mind that research indicates that women are more likely than men to adopt interpersonally oriented leadership styles.

The issues of discrimination not only affect their pay levels; women managers continue to experience direct discrimination in all aspects of their career development and progression. This discrimination frequently comes from the view that women do not possess the personal qualities required to be a good manager, and also from the assumption that they lack the commitment and motivation needed to succeed. Male managers are often seen as highly motivated and success-oriented individuals because they are constantly on the lookout for any means of progression, and will exploit every reasonable opportunity to get ahead (White, Cox and Cooper, 1992). In contrast, women are seen as only attempting to move up the career ladder if they feel competent to do so. This approach has been interpreted as a demonstration that women are unwilling to be recruited into executive positions (Wahl, 1995). Men are more likely than women to have negative views of women managers and to assign them with negative female traits (Deal and Stevenson, 1998). Consequently, it has often been contended that women fail to demonstrate as much interest in managerial careers as men, or have the real business understanding that is required for senior management. This assumed lack of interest has been consistently used to explain the under-representation of women in management, thereby placing the blame for their predicament firmly with women (Adler, 1993; Wahl, 1995). However, numerous studies have reported no differences regarding achievement motivation, aspirations towards promotion, or motivation to manage (Davidson and Cooper, 1992; Davidson and Burke, 1994) and some studies have found women managers to be more ambitious and committed to their careers than their male counterparts (Nicholson and West, 1988; Lahtinen and Wilson, 1994).

Organizational cultures

Organizational cultures, which are largely dominated by male values, are a major factor in the maintenance of the perception that women are not as ambitious or career-oriented as men and a significant barrier to women's career progression (Catalyst, 1991; Corcoran-Nates and Roberts, 1995). Numerous studies have illustrated that both male managers and management students believe that so-called 'masculine' traits are related to being a successful manager compared to 'feminine' traits, i.e. 'think manager – think male' (Schein and Davidson, 1993). In order to achieve success women typically have to adapt to the organizational culture by taking on male values and attitudes, but whilst this may appear to benefit the individual it leads to the overall marginalization of women. Research has shown that women who are allowed into these male domains often feel isolated and alone, and are unable to relate to either male or female colleagues (Marshall, 1995). In general, men are unwilling to develop a culture in which women can be included on more equal terms; they see it as a women's problem and unconnected to their own behaviour (Wahl, 1995). This means that women often feel that they are fighting a continual battle, with little or no support and little chance of change (Marshall, 1995).

Women have a strong need for achievement yet their motivation and career commitment remain under question as long as they retain the biological responsibilities for childbirth (White et al., 1992). Managerial career patterns are normatively predicted on the 'male model' of employment continuity and commitment. A male-dominated occupation does little to accommodate women who are simply expected to conform to 'masculine' norms in the pursuit of their careers. Women who deviate from this norm can be severely disadvantaged. For many women, bearing and rearing children results in a 'broken' or 'bimodal' career pattern during their late 20s and early 30s. This is a crucial time in career development when those with continuous career patterns are consolidating their positions (Spencer and Podmore, 1987). This situation has also resulted in the unfounded belief that women are 'quitters' who do not make the necessary investment in an organization to earn career opportunities, and thus do not merit the investment of the corporation (Brett and Stroh, 1994). This places women at an obvious disadvantage and women who have discontinuous work patterns are unfairly penalized by the majority of employers (Davidson, 1996). Interestingly, dilemmas related to whether to marry/live with someone and start a family, were one of the major stressors facing female managers found by Davidson and Cooper (1983, 1992).

Indirect exclusion

Career progression for managers may also be highly dependent on the composition and extent of their informal business networks. These networks can provide advice, support, knowledge, influence, information and sponsors (Burke, Rothstein and Bristor, 1995). The networks of male and female managers are substantially different, in both their composition and their degree of influence. Men's networks mainly consist of men and afford access to those who have influence over critical human resource decisions, such as promotion and recruitment. In contrast, women's networks contain a larger number of women, with fewer influential members (Burke et al.,, 1995; Ibarra, 1993). Previous research has consistently shown that managerial women are excluded from the business networks that are available to their male counterparts. This means that, because women managers do not have access to the same informal business networks as their male colleagues, they are denied the same information and assistance (Travers and Pemberton, 2000).

It has been suggested that there are several reasons why women lack access to these informal networks: (1) women may not be aware of the importance and potential usefulness of informal networks; (2) they are less skilled at playing the 'network game'; (3) women do not want to play the 'network game'; (4) men want to maintain male dominance by excluding women from informal interactions (Arroba and James, 1989; Burke et al., 1995). Whilst the first two points may influence the degree of women's integration into male-dominated networks, the third and fourth points seem to be the most critical factors. Women managers appear to be reluctant to engage in the politics that are an integral part of organizational life and business networks. Welsh (1980) believes that a lack of self-esteem is one of the factors that makes women so reticent to use business networks, although Arroba and James (1989) suggest that there is a more important factor that explains the reluctance of women managers to engage in politics. They propose that women managers frequently choose not to become involved in business networks because they view political activity with distaste, believing that they would be compromising their own principles if they were to enter into such relationships.

The reluctance of women managers to play the 'network game' is compounded by the tradition upheld by many male managers of 'looking after their own' (Davidson and Cooper, 1992). In situations, such as promotion, where male managers need assistance they turn to other men for assistance. As men hold the majority of senior management positions, these contacts are likely to provide valuable information and influence in relation to potential job opportunities. In contrast, women have to rely on a wider range of contacts because of the difficulty they face in finding suitable employment and will utilize any contact they can to find work, although these contacts may not be as effective as those available to men (Still and Guerin, 1986).

Direct exclusion

By any means of discrimination, exclusion frequently leads to feelings of isolation and despair, which is often exacerbated by lack of female role models in higher managerial positions. 'Token women' working in non-traditional jobs suffer most from stress that is related to discrimination and prejudice at work (Davidson and Fielden, 1999). This is particularly potent for black and ethnic minority managers, who are doubly disadvantaged in terms of upward mobility by high levels of work stress and pressure (Greenhaus, Parasuraman and Wormley, 1990; Hite, 1996; Bhavnani and Coyle, 2000). Davidson (1997) found that, although Black and ethnic women managers cited similar work stressors to white women managers, what made those stressors different was the double bind (i.e. racism and sexism). Bell and Nkomo, (2000) found that black professional and managerial women in America perceived themselves as living in a bi-cultural world, one white and one black. Consequently, they felt a constant 'push and pull' between different cultural contexts in their lives, resulting in high stress levels especially with regard to role conflict. In addition, James (1994) strongly emphasized the importance of social identities as an effective stress-coping mechanism. Thus, black and ethnic women managers not only face higher levels of stress, they are also frequently denied access to effective coping strategies specific to their minority status.

Isolation and 'tokenism' of any minority group can lead to bullying and harassment in the workplace. Sexual harassment is a pervasive form of victimization prevalent at all occupational levels, affecting millions of women each year (Bowes-Sperry and Tata, 1999). The figures relating to sexual harassment vary from country to country, reflecting both cultural and legal differences. However, the effects of sexual harassment are fairly common, with victims often experiencing adverse psychological, physical and behavioural outcomes. These include: depression, anger, anxiety, nausea, insomnia, increased alcohol consumption, as well as drug dependency (Terpstra and Baker, 1991; Wright and Bean, 1993).

Extra-organizational stressors

Extra-organizational stressors focus on the interface between work and home commitments, taking issues such as poor domestic support, dilemmas about starting a family, and so on (Davidson, 1997; Davidson and Fielden, 1999). These stressors are interactive and cumulative, forming an integrated whole that affects performance, behaviour, job satisfaction and well-being. Women managers continue to report greater levels of stress as a consequence of their experiences in and out of the working environment (Barnett and Brennan, 1997). High workloads, coupled with a greater responsibility for duties related to the home and family, mean that women managers are working far longer hours than their male counterparts. Long working hours are now recognized as posing a serious threat to health and well-being (Sparks and Cooper, 1997; Cooper, 1999), which means that women managers are inevitably facing increasing ill health as a result of their lifestyle. Many occupations are so demanding that they are considered to be a 'two-person' career, which implies that the back-up support of a spouse is essential. However, for women, marriage does not appear to have the same beneficial effects as it does for men (Schwartzberg and Dytell, 1989). Furthermore, women managers with children are potentially more at risk because of the higher stress levels they encounter (Nelson and Burke, 2000). In a study of dual career couples, Barnett and Shen (1997) found that distress in women was particularly linked to time spent in low-schedule control tasks. These were tasks commonly done under time pressure and with urgency, such as picking up children from childcarers, preparing meals, housecleaning, etc. Furthermore, lack of spousal support for female managers in relation to household duties has also been associated with poorer mental health (Beatty, 1996; Dunahoo, Geller and Hobfull, 1996).

The stress that arises from the conflicting nature of work and home pressures can also have a significant impact on career development and progression (Cooper and Lewis, 1998). For men, marriage provides a platform of support and security from which to launch their career, while for women it is a competing demand and an obstacle which represents a barrier to women's career progression (Cooper and Davidson, 1994; Greenglass, 1993). There is some evidence to suggest that both men and women are placing greater value on shorter working hours, in order that they may achieve a more balanced lifestyle (Cooper, 1999). Unfortunately, this attitude is seen by many organizations as reduced job commitment, rather than a reaction to the increasing demands of work and family life. Resistance to this type of change may arise from the unfounded belief that women would be the main beneficiaries of such change. Reduced working hours would place greater pressure on men to increase their participation in domestic roles. However, the work of Parasuraman, Greenhaus and Granrose (1992) indicates that men are already resistant to assisting in such roles, even when their partners experience high levels of work and family role stressors. It is perhaps not surprising that Galinsky, Bond and Friedman (1993) found that 60% of the women questioned had accepted their current position so as to minimize the negative effects on their personal and private lives. Nevertheless, this type of compromise can lead to increased levels of frustration at lost opportunities, even though the decision not to pursue such opportunities is recognized as lying with the individual (Gottlieb, Kelloway and Barham, 1998).

Individual personality characteristics

Self-concept

It is not only the sources of stress that contribute to the levels of stress experienced by women managers. The way in which individuals themselves react to those stressors also has a significant impact on the effect such stressors have on mental and physical well-being. Work is a highly valued activity and members of occupational groups, such as managers, often develop occupational self-images, which provide motivation and work satisfaction. According to Burns (1980) 'self-concept is a composite image of what we think we are, what we think others think of us and what we would like to be'. An individual's self-concept contains their experiences of their own body, their possessions, their family, their motive structure, drive status, defences and the feelings of pride and shame associated with these facets (Bala and Lakshmi, 1992). Positive self-images can maintain or increase psychological resistance to work-based pressures, protecting individuals from the adverse effects of stress (Kaufman, 1982). As women tend to hold less prestigious management positions than their male counterparts and experience lower levels of societal recognition (Alvesson, 1993), they are less likely to benefit from positive occupational self-images.

Self-esteem

Self-concept is intrinsically linked to an individual's level of self-esteem, which may be defined as the degree to which we like and value ourselves and operates at two levels (Jex, Cvetanovski and Allen, 1994). Self-esteem has two components, (1) relatively stable personality traits that relate to an individual's overall sense of self-esteem and develop global evaluations of their personal worth and competence (McCrae and Costa, 1990), and (2) domain-specific evaluations (e.g. job-related self-esteem), which fluctuate in response to the environment and variables specific to that environmental context (Pierce, Gardner, Dunham and Cummings, 1993). Individuals with low self-esteem are prone to doubt the efficacy and accuracy of their beliefs and behaviours, demonstrating greater responsivity to social cues and an increased desire to please (Jex et al., 1994). In addition they tend to report greater levels of anxiety and depressions than those with higher levels of self-esteem (Pierce et al., 1993).

In general, working women tend to hold more positive appraisals of others' views of them than working men, suggesting that they have higher levels of self-esteem (Bala and Lakshmi, 1992). However, because of the disparity between the managerial levels held by men and women, coupled with high levels of discrimination and exclusion, it is likely that the personal evaluations of women managers will be less favourable than those of men. Jex et al. (1994) propose that both global and domain-specific self-esteem moderate the relationship between stressors and stress outcomes. As women managers' access to this coping strategy is often restricted, their ability to protect themselves from the sources of stress they encounter will be seriously reduced.

Self-efficacy

Self-efficacy is defined as the belief in one's ability to perform a task, or more specifically to execute a specified behaviour successfully. Low self-efficacy interacts with low self-esteem to further reduce an individual's ability to deal effectively with workplace pressures. (Bandura, 1982). According to Bandura (1982) there are two types of expectancies that influence the choice of activities people will engage in, the amount of effort they will expend, and how long they will persist in the face of obstacles or aversive experiences. These are: outcome expectancy, the belief that certain behaviours will lead to certain outcomes, and self-efficacy expectancy, the belief that one can successfully perform the behaviours in question (Maddux, Sherer and Rogers, 1982). Those with a poor sense of self-efficacy will doubt their own capabilities and as these doubts grow they are likely to reduce their efforts or give up altogether, whereas those with a strong sense of self-efficacy will exert the greatest effort to master the challenges, maintaining high levels of performance (Bandura, 1982). As women frequently report lower levels of self-efficacy than their male counterparts (Vianen and Keizer, 1996), this situation has significant implications for the way in which women managers perceive pressure at work.

In relation to outcome and self-efficacy expectancies, individuals tend to make judgements about their own capabilities in two main ways, general-efficacy and domain-efficacy. General-efficacy is where an individual has a general set of expectations of their overall performance in a variety of situations, whereas domain-efficacy describes the beliefs about one's ability in a particular aspect of life (Woodruff and Cashman, 1993). It has been argued that women tend to have lower expectations of workplace achievement because they define objective success differently from men, rather than because of perceived differences in self-efficacy (Dann, 1995). Women managers may pursue positions at lower levels of management because they are judging themselves and their success by different standards to their male counterparts. Thus, rather than experiencing lower levels of domain-efficacy, they experience lower levels of general-efficacy which results in women imposing levels of achievement on themselves that do not coincide with those of the male approach to management (Dann, 1995).

Control

According to Rotter (1966), people have generalized expectancies as to whether or not they have the ability to control situations through their own actions. Individuals who do experience a general sense of personal control are considered to have an 'internal' locus of control, whereas those who do not have a general sense of personal control are considered to have an 'external' locus of control. 'Internal' locus of control is associated with many instrumental characteristics, such as assertiveness, independence, dominance and efficiency. Individuals with an internal locus of control tend to have a high need for achievement, a strong desire to assume personal responsibility for performing a task, take more initiatives in their efforts to attain their goals, seek high levels of information, and adopt behaviour patterns that facilitate personal control (Lefcourt,

1982; Kapalka and Lachenmeyer, 1988; Cherrington, 1991). They tend to perceive less stress, employ more task-centred coping behaviours and employ fewer emotion-centred behaviours than externals (Anderson, 1977). In general, people with an internal locus of control tend to develop fewer psychological disorders than those with an external locus of control (Weiten, 1989).

Previous research has shown that, in general, individuals employed in supervisory and management positions score higher on internal locus of control than those working in non-supervisory positions (Mellinger and Erdwins, 1985; Kapalka and Lachenmeyer, 1988; St-Yves, Contant, Freeston, Huard and Lemieux, 1989). Women in high level executive positions have been found to experience lower levels of strain than other women, regardless of their domestic situation, because they work in conditions of high control (Beatty, 1996). However, in general, women managers tend to hold lower managerial positions and are thus likely to experience a relatively low sense of personal control, compared to the small number of women in senior managerial positions. Consequently, compared to men managers as a whole, women managers tend to report lower levels of internal control and are more likely to employ emotion-centred behaviours, further increasing their chances of experiencing psychological ill health as a result of work stress (Vingerhoets and Van Heck, 1990; Hotchwarter, Perrewe and Dawkins, 1995).

Coping

Coping is generally defined as constantly changing cognitive and behavioural efforts to manage the internal and external demands of transactions that tax or exceed a person's resources (Latack, Kinicki and Prussia, 1995). It refers to the cognitive and behavioural efforts to manage the demands faced by an individual as a result of their situation. The process of coping with stressful events, such as work stress, is complex and highly dynamic and is directed toward moderating the impact of such events on an individual's physical, social and emotional functioning. The coping strategies adopted by an individual are determined by a number of factors including: personality variables (e.g. personal control), demographic factors (e.g. age and gender), sociodemographic factors (e.g. education and income) and availability of coping resources (e.g. self-esteem and experience) (Holahan and Moos, 1987; Gist and Mitchell, 1992). In addition, an individual's ability to cope with stress may be adversely affected by their tendency toward type A behaviour patterns which are often elicited by environmental stressors or challenges (Greenglass, 1993). Type A behaviour is characterized by extremes of competitiveness, striving for achievement, aggressiveness and haste, and has been found to be a significant source of stress-related illness in both female and male managers (Cooper and Davidson, 1982; Davidson and Cooper, 1983; Davidson et al., 1995).

It is common to distinguish between two major dimensions of coping: *problem-focused coping*, which addresses the stressful situation, and *emotion-focused coping*, which deals with the feelings and reactions to the stressful event (Latack et al., 1995). Problem-focused coping has been found to decrease emotional distress and is negatively related to depression, whereas emotional-focused coping increases emotional distress and is positively related to depression (Mitchell, Cronkite and Moos, 1983; Vitaliano, Maiuro, Russo and Becker, 1987). Vingerhoets and Van Heck (1990) found that men are more inclined to use active problem-focused coping strategies; they plan and rationalize their actions, they engage in positive thinking, perseverance, self-adaptation and personal growth. In contrast, women prefer emotional-focused solutions, they engage in self-blame and wishful thinking, they seek social support and a forum for the expression of their emotions. Where others are involved coping is often protracted and unpredictable, as the individual has to take account of other expectations when deciding how to handle the situation (Oakland and Ostell, 1996). Women are more likely to engage in behaviours that involve external recognition, allowing others to label and offer help with their problems, whereas men tend to deal with their problems internally (Astor-Dubin and Hammen, 1984). For female managers, working in male-dominated environments and cultures, this approach tends to be ineffective as it is in direct conflict with the values prevalent in such organizations. If women managers cannot successfully deal with this conflict they may be unable to find an effective means of coping with their situation, resulting in poorer psychological well-being, lower self-confidence and lower self-esteem (Holahan and Moos, 1987; Oakland and Ostell, 1996).

Stress outcomes

It is widely recognized that some stressors, e.g. managerial unemployment, result in impaired behavioural changes such as decreased job performance, job satisfaction and career aspirations, as well as lower levels of psychological and physical well-being in both men and women (Fielden and Davidson, 2000). There is, however, a great deal of conflict in the literature regarding the extent of the detrimental outcomes experienced by women. This conflict arises from outdated stereotypical views and a lack of understanding surrounding the importance of gender in explaining differences in stress outcomes (Walters, 1993; Davidson and Fielden, 1999). Jick and Mitz (1985) suggest that women experience psychological stress (e.g. depression, emotional discomfort) more frequently than men, whereas men experience physiological stress (e.g. coronary heart disease) more frequently than women. However, recent research has indicated that this latter belief is unfounded, and the evidence suggests that the links between stress and heart disease are now major concerns for both men and women (Kirtz-Silverstein, Wingard and Barrett-Connor, 1992; Elliott, 1995).

Gender differences have frequently been reported in relation to occupational stress and previous research has indicated that female managers react differently to male managers in terms of reported stress outcomes (Davidson et al., 1995). Stress-related illness tends to manifest itself in terms of physical ill health for male executives, whereas for female executives it is more likely to develop into mental ill health (Cooper and Melhuish, 1984). However, a recent study that used noradrenaline (norepinephrine)levels as a physiological measure of the stress response, found that women managers, especially those with children, had significantly higher noradrenaline levels than their male counterparts (Lundberg and Frankenhaeuser, 1999). This may arise for two reasons: firstly, poor mental health is not necessarily regarded as an illness and frequently 'genuine' sickness is only seen as being a physical condition, so male managers may consider physical illness more compatible with their self-concept and more acceptable in others' appraisals of them than they would mental illness (Miles, 1988). Secondly, women tend to normalize their mental health problems and therefore are far more likely to report higher levels of psychological distress (Walters, 1993).

Deaux (1984) suggests that an inherent difference in the mental health of men and women is not due to sex but is a reflection of the gender socialization process, and its role in influencing attitudes and behaviours. It has also been argued that these differences will continue as long as researchers continue to use a model of mental health that is based upon the assumption that 'normality = white, middle-class male', a model in comparison to which all other groups are still being judged (Frosh, 1987). Nelson and Quick (1985) maintain that women suffer from poorer mental health, not because they are inherently less stable than men, but because they experience greater sources of both psychological and physiological stress than men. Women are also more likely to experience psychosocial sources of stress than men, and a significant relationship between psychosocial stressors and susceptibility to infectious disease has been found (Arnetz, Wasserman, Petrini, Brenner, Levi, Eneroth, Salovaara, Hielm, Salovaara, Theorell and Petterson, 1987). In addition, recent research suggests that women's susceptibility to physiological impairment is increasing, with the links between stress and heart disease a major concern for both male and female managers (Elliott, 1995).

Conclusions

The research suggests that women managers, compared to those in clerical positions, are more satisfied and less stressed (Long, 1998). Yet the same appears to be true for all those in managerial positions compared to those in non-managerial positions, because of their access to greater control and resources (Kaufman, 1982; Beatty, 1996). Thus, this type of comparison is unhelpful in understanding the unique position of women managers.

It is only relatively recently that research into occupational stress and women in managerial positions has received any real attention. This research has clearly shown that, in comparison to their male counterparts, female managers are faced with many additional sources of stress arising as a direct result of discrimination and prejudice in the workplace (particularly in male-dominated industries) and increased home/work conflicts (Figure 7.1).

Women managers experience high levels of gender stereotyping, discrimination, direct and indirect exclusion, especially in organizational cultures that perpetuate the 'manager = white, middle-class, male' myth. They have to strive harder for less reward, both at work and in the domestic arena, with poorer accesses to effective coping strategies. There appears to be a substantial failure by organizations to recognize the difficulties women must overcome if they are to succeed in management, placing them under abnormally high levels of pressure. It is perhaps not surprising then that women managers tend to report significantly poorer mental well-being, characterized by poor self-image, low self-esteem, low self-efficacy and self-doubt.

References

Adler NJ (1993) An international perspective on the barriers to the advancement of women managers. Applied Psychology: An International Review 42(4): 289–300.

Alvesson M (1993) Cultural Perspectives on Organisations. Cambridge: Cambridge University Press.

Anderson CR (1977) Locus of control, coping behaviours, and performance in a stress setting: a longitudinal study. Journal of Applied Psychology 62(4): 446–51.

Arnetz BB, Wasserman J, Petrini B, Brenner SO, Levi L, Eneroth P, Salovaara H, Hielm R, Salovaara L, Theorell T, Petterson LL (1987) Immune function in unemployed women. Psychosomatic Medicine 49(1): 3–12.

Arroba T, James K (1989) Are politics palatable to women managers? How women can make wise moves at work. Women in Management Review 3(5): 123–30.

Astor-Dubin L, Hammen C (1984) Cognitive versus behavioural coping responses of men and women: a brief report. Cognitive Therapy and Research 8: 85–90.

Bala M, Lakshmi (1992) Perceived self in educated employed and educated unemployed women. International Journal of Social Psychiatry 38(4): 257–61.

Bandura A (1982) Self-efficacy mechanism in human agency. American Psychologist 37(2): 122–47.

Barnett RC, Brennan RT (1997) Changes in job conditions, changes in psychological distress and gender: A longitudinal study of dual-earner couples. Journal of Organizational Behavior 18(6): 459–78.

Barnett RC, Shen YC (1997) Gender, high -and low- schedule-control housework tasks and psychological distress: A study of dual earner couples. Journal of Family Issues 18: 403–28.

Beatty CA (1996) The stress of managerial and professional women: is the price too high? Journal of Organizational Behavior 17(3): 233–51.

Bell E, Nkomo S (2000) Refracted lives: Sources of discrimination between black and white women. In: Davidson MJ, Burke RJ (eds) Women in Management: Current Research Issues, vol. II. London: Sage.

Bhavnani R, Coyle A (2000) Black and minority ethnic women managers in the UK – Continuity or change? In: Davidson MJ, Burke RJ (eds) Women in Management: Current Research Issues, vol. II. London: Sage.

Bowes-Sperry L, Tata J (1999) A multi-perspective framework of sexual harassment. Reviewing two decades of research. In: Powell GN (ed.) Handbook of Gender and Work. London: Sage.

Burke RJ, Rothstein MG, Bristor JM (1995) Interpersonal networks of managerial and professional women and men: descriptive characteristics. Women in Management Review 10(1): 21–7.

Burns RB (1980) The Self-Concept Theory. London: Longman.

Cartwright S, Cooper CL (1997) Managing Workplace Stress. London: Sage.

Catalyst (1991) Women in corporate management: results of a catalyst survey. New York: Catalyst.

Catalyst (1996) Women in Corporate Leadership: Progress and Prospects. New York: Catalyst.

Cherrington DJ (1991) Needs theory of motivation. In: Steers RM, Porter LW (eds) Motivation and Work Behavior. New York: McGraw-Hill.

Cooper CL (1999) The changing psychological contract at work. European Business Journal 11(3): 115–18.

Cooper CL, Cooper CR, Eaker LH (1988) Living with Stress. London: Penguin.

Cooper CL, Davidson MJ (1982) High Pressure: Working Lives of Women Managers. London: Fontana.

Cooper CL, Davidson MJ (1994) (eds) Women in Management. London: Heinemann.

Cooper CL, Lewis S (1998) Balancing Your Career, Work and Family. London: Kogan Page.

Cooper CL, Marshall (1976) Occupational sources of stress: a review of the literature relating to coronary heart disease and mental ill-health. Journal of Occupational Psychology 49: 11–28.

Cooper CL, Melhuish A (1984) Executive stress and health: differences between men and women. Journal of Occupational Medicine 26(2): 99–103.

Corcoran-Nates Y, Roberts K (1995) 'We've got one of those': the peripheral status of women in male dominated industries. Gender, Work and Organisation 2(1): 21–33.

Dann S (1995) Gender differences in self-perceived success. Women in Management Review 10(8): 11–18.

Davidson MJ (1996) Women and employment. In: Warr P (ed.) Psychology at Work. London: Penguin.

Davidson MJ (1997) The Black and Ethnic Minority Woman Manager. London: Paul Chapman Publishing.

Davidson MJ, Burke RJ (eds) (1994) Women in Management - Current Research Issues. London: Paul Chapman.

Davidson MJ, Burke RJ (eds) (2000) Women in Management - Current Research Issues, vol. II. London: Sage.

Davidson MJ, Cooper CL (1983) Stress and the Woman Manager. London: Martin Robertson.

Davidson MJ, Cooper CL (1992) Shattering the Glass Ceiling: The Woman Manager. London: Paul Chapman Publishing.

Davidson MJ, Cooper CL, Baldini V (1995) Occupational stress in female and male graduate managers. Stress Medicine 11: 157–75.

Davidson MJ, Fielden SL (1999) Stress and the working woman. In: Powell GN (ed.) Handbook of Gender and Work. London: Sage.

Deal JJ, Stevenson MA (1998) Perceptions of female and male managers in the 1990s. Sex Roles: A Journal of Research 38: 287–301.

Dunahoo CL, Geller PA, Hobfull SE (1996) Women's coping: Communal versus individualistic orientation. In: Schabracq MJ, Winnubst JAM, Cooper CL(eds) Handbook of Work and Health Psychology. Chichester: Wiley.

Elkin A, Rosch P (1990) Promoting mental health at the workplace: The prevention side of stress management. Occupational Medicine: State of the Art Review 5: 739–54.

Elliott SJ (1995) Psychological stress, women and heart health: a critical review. Social Science and Medicine 40(1): 105–15.

Equal Opportunities Commission (1996) Equal Opportunities Review 65: 5–6.

Fielden SL, Davidson MJ (2000) Coping with Managerial Unemployment. London: BPS.

Frosh S (1987) The Politics of Psychoanalysis: An Introduction to Freudian and Post-Freudian Theory. London: Macmillan Educational.

Galinsky E, Bond JT, Friedman DE (1993) The National Study of the Changing Workforce. New York: Families and Work Institute.

Gardiner M, Tiggemann M (1999) Gender differences in leadership style, job stress and mental health in male-and-female dominated industries. Journal of Occupational and Organisational Psychology 72: 301–15.

Gist M, Mitchell T (1992) Self-efficacy: a theoretical analysis of its dimensions and malleability. Academy of Management Review 17: 183–211.

Gottlieb BH, Kelloway K, Barham E (1998) Flexible Work Arrangements: Managing the Work-Family Boundary. Chichester: John Wiley & Sons.

Greenglass ER (1993) Structural and social-psychological factors associated with job functioning by women managers. Psychological Reports 73(3): 979–86.

Greenhaus JH, Parasuraman S, Wormley WM (1990) Effects of race on organizational experiences, job performance, evaluation and career outcomes. Academy of Management Review 33(1): 66–86.

Hite LM (1996) Black women managers and administrators: experiences and implications. Women in Management Review 11(6): 11–17.

Holahan CJ, Moos RH (1987) Personal and contextual determinants of coping strategies. Journal of Personality and Social Psychology 52: 946–55.

Hotchwarter WA, Perrewe PL, Dawkins MC (1995) Gender differences in perceptions of stress-related variables: Do the people make the place or does the place make the people? Journal of Managerial Issues 7(1): 62–74.

Ibarra H (1993) Personal networks of women and minorities in management. Academy of Management Review 18: 56–87.

Institute of Management and Remuneration Economics (1998) UK National Management Survey. London: Institute of Management.

James K (1994) Social identity, work stress and minority workers health. In: Keita GP, Hurrell JJ (eds) Job Stress in a Changing Workforce, pp. 127–46. Washington DC: American Psychological Association.

Jex SM, Cvetanovski J, Allen SJ (1994) Self-esteem as a moderator of the impact of unemployment. Journal of Social Behavior and Personality 9(1): 69–80.

Jick TD, Mitz LF (1985) Sex differences in work stress. Academy of Management Review 10: 408–20.

Kapalka GM, Lachenmeyer JR (1988) Sex-role flexibility, locus of control, and occupational status. Sex Roles 19 (7 & 8): 417–27.

Kaufman HG (1982) Professionals in Search of Work: Coping with the Stress of Job Loss and Unemployment. New York: Wiley.

Kirtz-Silverstein D, Wingard DL, Barrett-Connor E (1992) Employment status and heart disease risk factors in middle-aged women: the Rancho Bernardo study. American Journal of Public Health 82(2): 215–19.

Lahtinen HK, Wilson FM (1994) Women and power in organisations. Executive Development 7(3): 16–23.

Latack JC, Kinicki AJ, Prussia GE (1995) An integrative process model of coping with job loss. Academy of Management Review 20(2): 311–42.

Lazarus RS (1971) The concept of stress and disease. In: Levi L (ed.) Society, Stress and Disease, vol. 1, pp 177–214. London: Oxford University Press.

Lefcourt HM (1982) Locus of Control: Current Trends in Theory and Research. Hillsdale, NJ: LEA.

Long B (1998) Coping with workplace stress: A multiple-group comparison of female managers and clerical workers. Journal of Counselling Psychology 45(1): 65–78.

Loscocco KA, Spitze G (1991) The organisational context of women's and men's pay satisfaction. Social Science Quarterly 72(1): 3–19.

Lundberg U, Frankenhaeuser M (1999) Stress and workload of men and women in high-ranking positions. Journal of Occupational Health Psychology 4(2): 142–51.

McCrae RR, Costa PT (1990) Personality in Adulthood. New York: Guilford Press.

Maddux J, Sherer M, Rogers R (1982) Self-efficacy expectancy and outcome expectancy. Cognitive Therapy and Research 6: 207–11.

Marshall J (1995) Working at senior management and board levels: some of the issues for women. Women in Management Review 10(3): 21–5.

Mellinger S, Erdwins P (1985) Personality correlates of age and life roles in adult women. Psychology of Women Quarterly 9: 503–14.

Miles A (1988) Women and Mental Illness. Brighton: Wheatsheaf.

Mitchell RE, Cronkite RC, Moos RH (1983) Stress, coping, and depression among married couples. Journal of Abnormal Psychology 92: 433–48.

Nelson DL, Burke RJ (2000) Women, work stress and health. In: Davidson MJ, Burke RJ (eds) Women in Management: Current Research Issues, vol. II, London: Sage.

Nelson DL, Quick JC (1985) Professional women: are distress and disease inevitable. Academy of Management Review 10: 206–13.

Nicholson N, West A (1988) Managerial Job Change: Men and Women in Transition. Cambridge: Cambridge University Press.

Oakland S, Ostell A (1996) Measuring coping: a review and critique. Human Relations 49(2): 133–55.

O'Driscoll MP, Cooper CL (1996) Sources and management of executive job stress and burnout. In: Warr P (ed.) Psychology at Work, pp. 188-223. London: Penguin.

Parasuraman S, Greenhaus JH, Granrose CS (1992) Role stressors, social support, and well-being among two-career couples. Journal of Organizational Behavior 13: 339–56.

Pierce JL, Gardner DG, Dunham RB, Cummings LL (1993) Moderation by organisation-based self-esteem of role condition-employee response relationship. Academy of Management Journal 36: 271–88.

Reskin BF, Ross CE (1992) Jobs, authority, and earnings among managers: the continuing significance of sex. Work and Occupations 19(4): 342–65.

Roos PA, Gatta, ML (1999) The gender gap in earnings: Trends, explanations and prospects. In: Powell GN (ed.) Handbook of Gender and Work. London: Sage.

Rosenfeld RA, Spenner KI (1992) Occupational sex segregation and women's early career job shifts. Work and Occupations 19(4): 424–49.

Rotter JB (1966) Generalised expectancies for internal versus external control of reinforcement. Psychological Monographs 80, Whole No. 609.

Schein VE, Davidson MJ (1993) Think manager, think male. Management Development Review 6: 24–8.

Schwartzberg NS, Dytell RS (1989) Family stress and psychological well-being among employed and non-employed mothers. Journal of Social Behaviour and Personality 3(4): 175–90.

Sparks K, Cooper CL (1997) The effects of hours of work on health: a meta-analysis review. Journal of Occupation and Organisational Psychology 70: 391–408.

Spencer A, Podmore D (eds)(1987) In a Man's World. London: Tavistock Publications.

Stevens CK, Bavetta AG, Gist ME (1993) Gender differences in the acquisition of salary negotiation skills: the role of goals, self-efficacy, and perceived control. Journal of Applied Psychology 78(5): 723–35.

Still LV, Guerin C (1986) Networking practices of men and women managers compared. Women in Management Review 2(2): 103–9.

St-Yves A, Contant F, Freeston MH, Huard J, Lemieux B (1989) Locus of control in women occupying middle-management and non-management positions. Psychological Reports 65: 483–6.

Terpstra DE, Baker DD (1991) Sexual harassment at work: the psychological issues. In: Davidson MJ, Earnshaw JM (eds) Vulnerable Workers, pp. 179–202. Chichester: John Wiley & Sons.

Travers C, Pemberton C (2000) Women's networking across boundaries – Key issues in networking within and between cultures. In: Davidson MJ, Burke RJ (eds) Women in Management - Current Research Issues, Vol. II. London: Sage.

Vianen AEM, Keizer WAJ (1996) Gender differences in managerial intention. Gender, Work and Organisation 3(2): 103–14.

Vingerhoets AJM, Van Heck GL (1990) Gender, coping and psychosomatic symptoms. Psychological Medicine 20: 125–35.

Vinnicombe S (2000) The position of women in management in Europe. In: Davidson MJ, Burke RJ (eds) Women in Management - Current Research Issues, vol. II. London: Sage.

Vitaliano PP, Maiuro RD, Russo J, Becker J (1987) Raw versus relative scores in the assessment of coping strategies. Journal of Behavioural Medicine 10: 1–18.

Wahl A (ed.) (1995) Men's Perceptions of Women and Management. Stockholm: Norstedts Tryckeri.

Walters V (1993) Stress, anxiety and depression: women's accounts of their health problems. Social Science and Medicine 36(4): 393–402.

Welsh MS (1980) Networking, the Great New Way for Women to Get Ahead? London: Sage.

White B, Cox C, Cooper CL (1992) Women's Career Development: A study of High Flyers. Oxford: Blackwell.

Woodruff SL, Cashman JF (1993) Task, domain, and general efficacy: a reexamination of the self-efficacy scale. Psychological Reports 72: 423–32.

Wright PC, Bean SA (1993) Sexual harassment: An issue of employee effectiveness. Journal of Management Psychology 8(2): 30–6.

Chapter 8
Stress in Teaching: Past, Present and Future

Cheryl J. Travers

> A primary school teacher hanged herself at home after she became terrified by a
> visit from OFSTED inspectors ... She worried about failing the inspection and
> the effect it might have on her career. (*The Express*, 30 September 1999)

Over the last 30 or so years it has been widely acknowledged that teaching
is a profession in crisis with members who are highly stressed. The
spotlight on this particular service provider has largely been due to the
vast amount of research on the topic and also the lamenting of teachers
themselves through their unions.

Traditionally researchers attempting to obtain information about the
extent of the problem have been faced with difficulties as victims have
feared that to report stress may be taken as a sign of weakness (Dunham,
1992). However, the sheer volume of changes in education legislation
(Brown and Ralph, 1997) in terms of the content required and methods
employed, alongside greater accountability and public assessment, have
given rise to new skill requisites and the resultant disillusioned attitudes
that teachers now have towards the job of teaching.

In addition, the recent landmark case of Walker versus
Northumberland County Council in November 1994 has opened the
floodgates for similar cases within the public sector, especially teaching.

This chapter will attempt to summarize the research in this area by first
defining what we mean by the term, and then by examining the costs of
teacher stress, followed by the causes. The role of the teacher in the stress
experience will also be examined and the chapter will conclude with some
suggestions for future research.

What do we mean by teacher stress?

Since stress is now recognized as the 'occupational disease with the most
significant ramifications for teachers and their employers and indeed for

lawyers at work, as employers and employees' (Usher, 1998), we need to be clear about our operational definition. An early offering by Kyriacou and Sutcliffe (1979a) suggested that teacher stress is:

> a response syndrome of negative affect (such as anger and depression), usually accompanied by potentially pathogenic physiological changes (such as increased heart rate) resulting from aspects of the teacher's job and mediated by the perception that the demands made upon the teacher constitute a threat to his (or her) self-esteem or well-being and by coping mechanisms activated to reduce threat.

This highlights that the extent and type of stress experienced by teachers will largely depend upon whether or not the teachers feel threatened by particular demands facing them, and the individual teacher may, after facing an initial threat, be able to modify or ameliorate the threat by particular actions.

Kyriacou (1998) suggests more recently that in attempting to produce a definition we need to consider whether the term is referring to (a) the level of demands made on the teacher (either positive or negative or both) and/or (b) the emotional states engendered in a person in attempting to meet such demands (again either positive or negative or both). Tending to focus on the negative aspects of teacher stress Kyriacou states that teacher stress is:

> the experience by a teacher of unpleasant emotions such as tension, frustration, anxiety, anger and depression, resulting from aspects of his or her work as a teacher (Kyriacou, 1997, p. 156).

It is important to recognize that we must consider the interaction between the objective nature of the situation and individual teacher's subjective appraisal of it – i.e. following the transactional models of Lazarus (1966) and Lazarus and Folkman (1984). At the same time we need to observe the balance between the actual demands placed on teachers and their ability to meet such demands. This is even more important in the light of recent litigation and teacher stress claims for compensation reported in the teacher press.

What is the extent of teacher stress?

Estimates of the percentage of teachers actually experiencing high levels of perceived stress have varied considerably over the years. Earlier reports suggested that the figure could be between 30% and 90% (Hawkes and Dedrick, 1983; Laughlin, 1984) and British research has revealed that between one-fifth and one-third of teachers report experiencing a great deal of stress (Pratt, 1978; Dunham, 1983). Travers and Cooper (1993) revealed that a quarter to one-fifth of teachers were self-reporting levels of mental ill health (depression, psychosomatic anxiety and free-floating anxiety) similar to that of psychoneurotic outpatients using the Crown–Crisp Experiential Health Index (CCEI; Crown and Crisp, 1979). However, a major breakthrough in estimating the extent of the problem has resulted from research comparisons between teachers and other occupational groups traditionally viewed as stressed. Using similar measures of stress-related symptoms research has found that teachers are amongst those groups displaying the highest levels of stress (e.g. Travers and Cooper, 1993).

Teacher stress is not just a British phenomenon, however. In an international review of teacher stress and burnout, Kyriacou (1987) referred to the occurrence and consequences of stress in the teaching profession in countries as widespread as Great Britain, the United States, Israel, Canada and New Zealand (e.g. Coates and Thoreson, 1976). More recently studies have examined teacher stress in Saudi Arabia (Milaat, 1997) and Hong Kong (Chan, 1998), and indeed the problem has been so well acknowledged in Holland due to government changes that in October 1997 they had 'World Teachers Day' (TES, 1997b). Increasingly attempts have been made to make cross-cultural comparisons (e.g. Travers, 1997; Pithers and Soden, 1998a) though similarities have tended to outweigh the differences found. The result is that it is now acknowledged that teacher stress is a worldwide problem.

How do teachers actually respond to stress?

A number of stress symptoms and responses have been observed in teachers over the years and what follows is a summary of the key findings.

Psychological responses to teacher stress

Mental ill health and teachers

Recently the National Association of School Masters and Women Teachers (NASUWT) have reported an increasing number of calls at the end of the summer holidays from teachers who cannot face returning to the classroom. Depression, sleepless nights and a breakdown in their relationships as a result of stress have been cited as the cause (TES, 1997b).

A number of research studies have highlighted the positive relationship between self-reported teacher stress and overall measures of mental ill health (e.g. Pratt, 1978; Galloway, Pankhurst, Boswell, Boswell and Green, 1982; Tellenback, Brenner and Lofgren, 1983).

Early work by Dunham (1976) identified frustration and anxiety as the two most common types of reactions to teacher stress. Frustration can be seen to be associated with the physiological symptoms of headaches, sleep disturbances, stomach upsets, hypertension and body rashes and in severe cases, depressive illness, whereas anxiety can be linked to loss of confidence, feelings of inadequacy, confusion in thinking and sometimes panic. In severe cases, anxiety can lead to the physiological psychosomatic symptoms of a nervous rash, twitchy eye, loss of voice and weight loss. In prolonged cases, a nervous breakdown or complete burnout may result.

Burnout in teachers – an extreme reaction to stress

Increasingly there has been a growing alarm at the rate of teacher burnout and the adverse implications this has for the learning environment in schools and on the achievement of educational goals. Referring to total emotional exhaustion (Hargreaves, 1978), this state of 'burnout' may lead to out-of-school apathy, alienation from work and withdrawal into a number of defensive strategies. Of major concern to the teaching profession is that 'burnout' can detract from the quality of teaching. Mancini, Wuest, Vantine and Clark (1984) have shown that 'burned-out' teachers give significantly less information and less praise, show less acceptance of their pupils' ideas and interact less frequently with them.

Pierce and Molloy (1990, p. 330) describe three aspects of burnout: 'the first is the development of increased feelings of emotional exhaustion and fatigue. Second is the tendency for teachers to develop negative cynical attitudes towards their students. The third aspect of burnout is the tendency to evaluate oneself negatively, resulting in feelings of lack of personal accomplishment.'

What causes teachers to burn out?

Studies into teacher burnout have shown that it is largely a result of excessive work stress over extended periods of time (Blase, 1982), and relentless work demands (Begley, 1982). A study of 33 teachers of emotionally disturbed children by Lawrenson and McKinnon (1982) revealed that a way of preventing burnout was to be aware of the stressful nature of the job.

A teacher's personality can also play its part. Nagy (1982) found that type A personality, workaholism and perceptions of working environment were individual factors that contributed to burnout. However, none of these were good predictors of its occurrence.

Teachers with a negative attitude towards students, an external locus of control (see later sections of this chapter) and intolerance of ambiguity are reported to have highest levels of burnout (Fielding, 1982). A further finding was that a school with a negative work climate exhibited a greater 'burnout-personality' relationship, than one with a positive work climate. A study of 100 teachers in the US by Zabel and Zabel (1982) revealed that young, less experienced teachers exhibited higher levels of burnout and yet less burnout was experienced by those receiving more support from administrators, fellow teachers and parents.

Westerhouse (1979) and Schwab (1981) have shown that role conflict and role ambiguity are significantly related to teacher burnout. A study of 40 American teachers by Cooley and Laviki (1981) concluded that individual, social-psychological and organizational factors were all strongly associated with the burnout response, and that it was important, therefore, to study all of these factors together to be able to understand the relative importance of these factors. Lowenstein (1991) revealed burnout to be a product of a lack of social recognition, large class sizes, lack of resources, isolation, fear of violence, role ambiguity, limited professional opportunities and lack of support. These are all stress factors reported by Britain's teachers (Travers and Cooper, 1993).

Job dissatisfaction

Travers and Cooper (1993) found that teachers' job satisfaction was significantly lower than other comparable occupational groups (i.e. doctors, nurses, tax officers) and that the major predictor of this was the pressure they experience from the management and structure of the schools in which they teach. Also, the lack of recognition that teachers are currently perceiving has a part to play in what is acknowledged as an alarmingly low level of job satisfaction.

One of the complexities of teaching as a profession, however, is that teachers can be suffering from occupational stress, but still gain job satisfaction from certain aspects of the job. Indeed Hart (1994) has revealed that if we reduce negative experiences in the teaching environment this will not necessarily enhance satisfaction. Also, if we enhance the positive aspects this may lead to greater job satisfaction, but not necessarily reduce psychological distress – i.e. they are not mutually exclusive.

More detailed analysis of the issues relating to this job dissatisfaction reveals that factors such as salary, career structure, promotion opportunities and occupational status are involved (Tellenback et al., 1983). Kyriacou and Sutcliffe (1979a) in a study of 218 teachers from mixed comprehensive schools in England found that self-reported teacher stress was negatively correlated with job satisfaction. However, they found that there was no significant difference in terms of age, length of experience and position held in school.

Needle, Griffin, Svendsen and Berney (1980) also found that teachers reporting higher levels of job stress reported greater job dissatisfaction. Kyriacou and Sutcliffe (1979a) found that job satisfaction was significantly negatively correlated to the following job stressors: poor career structure; individual misbehaving pupils; inadequate salary; inadequate disciplinary policy of school; noisy pupils; difficult classes; trying to uphold/maintain standards; too much work to do.

In addition, other studies have discovered that older, more experienced teachers tend to express higher levels of job satisfaction. Chaplain's study of 267 primary teachers also revealed younger (under 35) and older teachers (over 45) were more satisfied than the intermediate group. Avi Itzhah's study of 93 female kindergarten teachers in Israel also found that age and teaching experience were positively related to satisfaction (TES, 1998b).

Behavioural responses to stress

Though there is little research scrutiny of the link between palliative coping and stress in teachers (e.g. smoking, drinking and drugs), we may suppose that teachers will be as vulnerable to these coping strategies as any other occupational group.

Travers and Cooper's (1993) study of UK teachers found that a large proportion of the sample were drinking above the recommended weekly average. In addition, a longitudinal study of London teachers revealed that the presence of alcohol indicators in their blood increased as the term progressed (Travers and Cooper, 1994).

Reports in the teacher press (e.g. TES, 1999a) hint that being in school increases smoking behaviour. With 63% of schools in the UK allowing smoking on the premises, reports reveal that many teachers say that they consume a greater number of cigarettes when at school than elsewhere.

Other areas of concern are the use of drugs to alleviate stress symptoms. A Hampshire survey (cited TES, 1997c) of 180 head teachers revealed that 40% were affected by stress, causing symptoms such as irritability and exhaustion. One in ten were taking prescribed anti-depressants or sedatives – most commonly Prozac, the anti-depressant medication.

Inadequate diet is another area that needs to be considered as a response to stress. Research (cited TES, 1998a) has revealed that unhealthy changes in food and alcohol consumption are caused by day-to-day hassles as well as major hassles. Steptoe and Lipsey of St George's Hospital Medical School in London asked both nurses and teachers to fill in daily assessments of mood, alcohol and food intake together with weekly measures of hassles, stress, anxiety, depression and exercise over an eight week period. They found that respondents' fast food consumption (especially cheese) was linked to high stress weeks. Also those who usually drink to cope with stress consume more at these times.

Withdrawal from teaching as a response to stress

Another set of symptoms associated with teacher stress are turnover, early retirement, sickness absenteeism and intention to leave – all forms of withdrawal. These are perhaps the options teachers take when they find themselves in intolerably stressful situations.

Turnover and early retirement

A turnover rate in any profession of between 7 and 8% may be seen as healthy, but in teaching this has been reported to be far greater. The resignations would appear to be affected by the subject areas in which teachers teach, the type of school and the sector. In a study by local authority employers and teacher unions (cited *The Independent*, 8 September 1990), it was revealed that higher rates of resignation were found in foreign language, business, commercial and music teachers. Other findings suggest that Greater London has been worse hit, and evidence reveals that teachers within the primary sector may be the most likely to 'escape' from the profession. This results in an unexpectedly older workforce in primary schools. The same survey discovered that half of all primary teachers were over 40 and very few are under 30. Recent findings reveal that 38% of teachers quitting teaching do so due to ill health compared to only 9% of nurses (TES, 1997a).

In addition to the problems of absenteeism and turnover, a large number of teachers are looking for early retirement as a way out of teaching. This is not to say that for the vast majority this is not legitimate on the grounds of ill health, but for many this is the only way they see to get away from the job that is causing them excessive pressure. This means the education system and society as a whole are losing a large proportion of its experienced workforce. Many have explained that this desire to leave early is indeed a reaction to the stress of the job.

Sickness absence in teachers

Simpson (1976) suggested that sickness absence is a way that teachers can allow themselves time to temporarily withdraw from stress at work, without having to make a definite break. It is believed that this allows teachers to continually readjust to stressful work situations by such occasional withdrawals, and at the same time, develop skills necessary to deal with the sources of stress that they face. A problem with this interpretation, however, lies with the fact that it is difficult to distinguish between somewhat 'voluntary' absenteeism related to psychological causes (e.g. depression) and stress-related physical illness. Kyriacou and Sutcliffe (1979a) in a study of 218 UK secondary teachers found an association between self-reported teacher stress, job satisfaction, absenteeism and intention to leave the teaching profession

Travers and Cooper (1993) found that absenteeism in UK teachers was largely due to stress-related causes (e.g. persistent virus, anxiety and depression, bowel and stomach disorders).

A more recent Labour Force Survey reveals that sickness absence rates are approximately 3.5% for teachers (depending upon the type of school) compared to 2.9% for other professional workers. Overall an increase in sickness absence in the public sector has been observed, with on average 7 days off for teachers compared to 12 for both police and prison officers. The teacher supply agency Capstan says that 35 000 vacancies per day due to illness are required to be filled in schools (TES, 1998d).

Intention to leave the profession

Almost as harmful as actual turnover is the number of teachers within the profession who would rather be out of it. Travers and Cooper (1993) found that 66% of their sample of UK teachers had actively considered leaving the profession in the five years prior to the survey. Some intentions do not come to fruition and some resignations are of an impulsive nature (Mobley, 1982). However, as teacher turnover intentions appear to be on the increase, it is important to be able to understand what factors have the most disruptive impact.

The most frequently cited predictors of withdrawal in general have been those of intrinsic and extrinsic rewards (Bridges, 1980). Studies into a teacher's intention to leave have come to differing conclusions. Intrinsic rewards (i.e. recognition, sense of accomplishment, fulfilment, advancement) have been found by some to play a more important role than extrinsic rewards (i.e. working conditions, management policies) in the process of withdrawal. As teachers are in the service sector, motivation is assumed to be linked with intrinsic rather than extrinsic rewards (Spuck, 1977). Size of classes, administrative and teaching loads, availability of teaching aids, social and work relations have all been found to be related to teachers' affective reactions, stress and turnover intentions (D'Arienzo, Moracco and Krajewski, 1982).

These turnover intentions have been found to be exacerbated by structural characteristics of the job, lack of extrinsic and intrinsic rewards but reduced by group support, social co-operation and good work relations (e.g. Golembiewski, Munzenrider and Carter, 1983).

In addition the influence of other stress reactions cannot be ignored. Studies have suggested that burnout leads to turnover intentions and affects withdrawal behaviours (e.g. Burke, Shearer and Deszca, 1984). Travers and Cooper (1993) found that reported poor mental ill health was by far the greatest predictor of intention to leave teaching. This could be interpreted as teachers intending to 'get out' in a final act of self-preservation and awareness.

Research has also attempted to explain which factors will inhibit teachers' intention to leave. The longer a teacher holds a particular job or is employed within a particular school, the more benefits and privileges they accrue which are not transferable. Because of this, some individuals will be psychologically constrained from leaving. Steers and Mowday (1981) suggest that certain aspects will enhance a teacher's choice to stay, i.e. individually tailored work conditions; financial rewards (e.g. pension plans); specialized information and skills; familiarity with organizational work procedures; seniority privileges; personal reputation, social standing or power. Specific restraining factors are the teacher's status within the particular school and the specificity of teacher training that they have received.

Causes of stress in teaching

It must be recognized that the sources of stress in teachers are multidimensional (Borg and Riding, 1991; Borg, Riding and Falzon, 1991) but the following sections will describe what are seen to be the main sources.

Physical working conditions

A large number of teachers in our society today find themselves faced by circumstances which they believe force them to do their job badly (Esteve, 1989), in particular poor physical working conditions (e.g. Wanberg, 1984), inadequate school buildings and equipment (e.g. Smith and Cline, 1980), an unpleasant work environment, class sizes and noise levels (Kyriacou and Sutcliffe, 1978b; Connors, 1983; Fimian and Santoro, 1983). These poor conditions are largely reinforced by a lack of resources (Laughlin, 1984). Esteve refers to these as 'primary factors', because they directly affect teaching, create limitations or produce tension in the teacher's day-to-day practice.

The general feeling among many teachers is that there is an apparent contradiction imposed upon them by outside bodies (i.e. the demand for modern methods but without the adequate equipment to do the job). This situation is exacerbated by reduced expenditure on equipment due to a worsening financial situation in a number of schools (Fimian and Santoro,

1983). With reorganization in education, a large number of schools are experiencing the 'split-site' phenomenon – this often means that teachers have to travel between two sites, which can have many time and practical implications.

Limitations in the working environment in which teachers work need not just be physical ones. Often institutional limitations are imposed upon their work. Goble and Porter (1980) and Bayer and Chauvet (1980) emphasize that the institutional framework within which they work often dictates what teachers can do (e.g. timetable problems, internal rules, standards that have been laid down by the inspecting bodies or teaching institutions). They are also often required to set time aside for staff meetings, students, governing bodies and examination meetings and parents evenings etc.

Teachers' workload as a source of stress

This has become increasingly important recently due to the recent litigation cases of teachers claiming to be forced to work excessive hours leading to stress.

Another aspect of the teaching profession, which may be directly related to work overload, is the problem of having a wide range of pupil abilities in one class. This may require more lesson planning and more detailed and lengthy assessment (Dunham, 1980; Fimian and Santoro, 1983; Hawkes and Dedrick, 1983). In addition, it may be linked to a poor teacher–pupil ratio (e.g. Kalker, 1984; Russell, Altmaier and Van Velsen, 1987).

Work overload is also heavily linked to time pressures, not only in terms of the amount of work teachers have to fit in during the day, but also the amount that they have to take home at night and weekends, intruding into their personal life (Smith and Cline, 1980; Austin, 1981; ILO, 1981; Fimian and Santoro, 1983). Many teachers claim to put in excessive hours at home marking, preparing and assessing work. This has been a major problem for primary teachers since the introduction of compulsory testing for seven-year-olds, for example.

Researchers have discussed the actual stress of the school day in terms of the constant workload that it imposes upon the teacher. In addition, Kyriacou (1997) has suggested that one of the main sources of stress for teachers may be the 'general level of alertness and vigilance required' of them. This 'pace' of the school day is perhaps more of a problem, because of its 'rigid' nature, i.e. the way it is structured and the fact that teachers spend so much of their day in direct contact with pupils and that they very rarely get a clear break for lunch, for example.

To a great extent, however, the teaching workload is very much dependent upon the time of year (e.g. end of term/year examinations), and research has attempted to examine this.

For example, a Canadian study by Hembling and Gilliland (1981) of teachers' experience of stress over a 12 month period revealed that the highest incidence of stress occurred at the end of each term and at the end of the school year. The researchers explained that this was due to the accumulated tension during the previous term and specific 'end of term' events (e.g. more often than not examinations are set at the end of term). The study also showed how the levels of stress could accumulate, emphasizing the importance of school holidays as a means of regaining personal stability.

Further studies have also utilized longitudinal methods. In a US study, Fleishut (1985) studied stress patterns in 81 elementary school teachers in Pennsylvania over 12 months. The study revealed that stress increased during the first part of the school year (September to November), and decreased after the Christmas holidays (in January), with a steady increase throughout March, and another high point experienced in May. But a study by the New York State United Teachers (NYSUT) (1980) found that the opening week of school was the most stressful time.

Travers and Cooper (1994) collected blood samples and questionnaires from inner city London school teachers on two occasions throughout the autumn term – at the start of term and at the end. They found that teachers were revealing low levels of cortisol, which suggested chronic fatigue, even at the start of term. This study also revealed that absenteeism data may not always be an accurate measure of teacher stress as many claimed that they never took time off during term time even when sick because of the risk of causing problems for other staff. It may be more effective to ask teachers how many days they actually turn up for work 'ill'.

Teachers' role in the school as a source of stress

Role conflict and role ambiguity

Change may lead to stress as it can introduce conflict or ambiguity into what was originally a stable teaching role (Kelly, 1974). However, Dunham (1984) has pointed out that change might equally be welcomed as an alleviation of stress, depending upon circumstances and participants. He studied the stress imposed by the demands of specific managerial roles in teaching and found that tension was created by role conflict and role ambiguity.

A large-scale study by Crane and Iwanicki (1986) investigated 443 urban, remedial education teachers in an attempt to discover a possible relationship between burnout, role conflict and role ambiguity. The picture they found was a confusing one with regard to the 'overall' moderating effect of burnout, which varied according to a complex interaction of age, sex, experience and setting.

There are a number of situations that may lead to role ambiguity that are contemporary issues in teaching, i.e. job relocation, changes in the method of working, new organizational structure and changes in actual requirements of the job. Role ambiguity is a pervasive part of the teacher's experience due to the endemic uncertainty regarding the teacher's role in the school (Schwab and Iwanicki, 1982; Bacharach, Bauer and Conley, 1986; Travers and Cooper, 1993).

The issue of role conflict may be seen to be very relevant to teachers as it may include both intra-role conflict due to contradictory expectations from parents, pupils, principals etc, and inter-role conflict due to teachers having to assume several roles within the school setting. The multiplicity of roles that the teacher may have to fulfil can include that of a diagnostician, guidance counsellor, remediator, evaluator and then finally teacher. Increasingly, the role of 'social worker' is becoming part of the teacher role (e.g. Phillips and Lee, 1980; Austin, 1981; Sparks and Hammond, 1981).

Sometimes role conflict may require teachers to reject their own principles and better judgement (Dunham, 1980). For example, due to staff shortages, they may be forced to teach a subject outside of their own special area and one for which they have no desire or skill (Burke and Dunham, 1982; Schwab and Iwanicki, 1982; Kalker, 1984). They may also have to spend a considerable amount of time controlling pupils and dealing with discipline problems at the cost of time spent on actual teaching (Kalker, 1984). Other problems that may create role conflict day-to-day are maintaining self-control when angry (New York State United Teachers (NYSUT), 1980) and finding it difficult to accept the limits of what teaching can achieve (Instructor, 1979).

The problems resulting from the occurrence of role conflict can be exacerbated when certain psychological processes of the individual come into play. For example, if their belief system emphasizes perfectionism and compulsive behaviour, this may result in excessive worrying about the situation and anticipation of the problems that result if they do not meet expectations (Moracco, Gray and D'arienzo, 1981). Clagget (1980) also emphasizes the problem of conflicting values as a source of role conflict. This may be very important for teachers at present, as they may not agree with a number of changes that have taken place within education (e.g. the content of the National Curriculum).

Role overload

When referring to role overload one is portraying a situation where too little time is available to devote to the teaching function (Needle, Griffin and Svendsen, 1981). This may be a concern of many contemporary teachers as more and more time is required to be spent on administrative and pastoral tasks and responsibilities. Problems connected with role overload may include constant interaction with pupils, which allows little time for relaxation, lunch etc. (Weiskopf, 1980), constant interaction with others (Schwab, 1983), too many roles altogether (Austin, 1981), and the problem of being physically and emotionally drained (Sparks, 1979).

Role preparedness

Another potential stressor is that of being inadequately prepared for the role of 'teacher', i.e. by inadequate training (Fimian and Santoro, 1983). With the rapid changes that have taken place within teaching in the last ten years, it is very possible that teacher training may well be out of date by the time they start to actually teach. Also it might well be that the 'teacher of today' has a very different role to that of a teacher starting a career ten years ago, and many teachers may find that it is not the job that they expected. This may also apply to teachers that have been in the profession for a number of years.

Cains and Brown (1998) reveal that newly qualified primary school teachers are more likely to feel that they had been effectively prepared for teaching than do their secondary counterparts.

The Role of senior managers in teaching

More recently, studies have addressed in more detail the effects on the head teachers, who are finding themselves in the dual position of being both a people manager and a financial manager.

A study by Cooper and Kelly (1993) assessed occupational stress amongst 2638 head teachers of primary and secondary schools, together with principals/directors of further and higher education establishments, throughout the UK. Data were collected on personal/job demographics, sources of job stress, mental health, job satisfaction and coping strategies. It was found that the levels of job dissatisfaction and mental ill health were higher in secondary and primary school teachers than those in the further/higher education (FHE) sector. In addition, it was found that female head teachers and principals in secondary and FHE seem to be suffering significantly greater job dissatisfaction than their male counterparts, although this does not translate itself into mental ill health. Male head teachers, on the other hand, seem to suffer more mental ill health than their female counterparts. And finally, the two main sources of occupational stress which are predictors of job dissatisfaction and mental ill health are 'work overload' and 'handling relationships with staff'.

Recent work by Ostell and Oakland (1999) suggests that heads may be seen to solve problems in terms of absolutist and non-absolutist thinking and that this can explain differences in their behaviour, emotion management and psychological health, so emphasizing the individual style of head teachers and their stress experience.

Relationships at work as a source of teacher stress

Much research has also revealed that teachers are experiencing stress from their relationships with: fellow teaching colleagues (Wanberg, 1984; Brenner, Sorbom and Wallius, 1985), head teachers (Clark, 1980; Needle et al., 1980; Tellenback et al., 1983), administrators/education authorities (Hawkes and Dedrick, 1983; Kalker, 1984), parents (Kalker, 1984; Mykeltun, 1984), the community (Cox, 1977) and pupils (Tellenback et al., 1983; Brenner et al., 1985).

Relationships with colleagues

Dunham (1977) found that working relationships with colleagues were reported as a source of stress for teachers. It has been argued that the dominant source of stress is the *quality* of these interpersonal relationships, and that good social relationships are of great value when providing support which may alleviate stress (Brenner et al., 1985).

Kyriacou (1981) has suggested that schools should attempt to improve the social support received by staff, and that a great deal of the responsibility for doing this must lie with the head teacher. It must be noted at this point, however, that good working relationships may only flourish if the organizational structure is designed in such a way that it facilitates good working relationships between individuals.

Another problem that may face teachers in schools is that they may fear protesting about their problems, when they are overburdened, because they do not want to let fellow teachers down. For example, although the only way to cope with stress might be absenteeism, they fear the resulting overload this may impose on other teachers in the school. Kyriacou (1987) explains that although absenteeism may enable some teachers to cope, it may have a resulting negative impact, as it can worsen relationships when the classes of an absent teacher have to be covered by others on the staff.

It might be possible to witness the development of 'factions' within schools. According to Claxton (1988), this may be one way in which the breakdown of relationships manifests itself in stressed organizations. In addition, Spanoil and Casputo (1979) stressed the development of a lack of trust in stressed organizations. The problem is that these 'divisions', although a manifestation of stress, can subsequently become a cause of stress themselves.

Relationships with pupils

One of the most potentially exhausting aspects of teaching must surely be the fact that teachers are, in most cases, constantly responsible for others (Weiskopf, 1980; Brenner et al., 1985).

Relationships with pupils have been suggested as the most important source of stress for teachers (Tellenback et al., 1983), Several studies have indicated that disruptive behaviour is consistently a predictor of teacher stress (Borg and Riding, 1991; Borg et al, 1991).

Dealing with pupils is the major aspect of their job, and many problems and potential sources of stress can result from this, as pupils can be disruptive, undisciplined and unpredictable in their behaviour. Teachers can, depending on the nature of the pupils, spend vast amounts of time controlling this poor behaviour (e.g. Mykeltun, 1984; Hawkes and Dedrick, 1983; Kalker, 1984; Laughlin, 1984; Wanberg, 1984; Russell et al., 1987) and also it can affect their feelings of self-confidence (Dunham, 1992). Also, according to Claggett (1980), many teachers try to live up to the 'good shepherd' ethic, whereby each teacher tries to ensure that each child is successful in school by providing for individual needs.

However, other studies have suggested contrary findings (e.g. Litt and Turk, 1985). There are various possible explanations as to why these contradictions exist. Teachers may actually differ in their willingness to admit to experiencing problems with pupils, as this is seen by many to be a major feature of the teacher's job. In addition, there are many different types and levels of misbehaviour, from minor examples of restlessness to serious physical attacks. When pupil misbehaviour has been examined in relation to stress, some studies have made no distinction between different types of behaviour problem, while others have concentrated solely on major stressful events (e.g. Comber and Whitefield, 1979). Whatever the findings, it has been suggested that single, serious disruptive incidents may be a lesser cause of stress than the cumulative effect of constant or repeated 'low level' disruption (Kyriacou, 1987). In addition, teachers have differing perspectives as to what constitutes a discipline problem.

Lack of pupil motivation may also be a source of stress for teachers. Kyriacou and Roe (1988) found that 'under-achieving' was rated as the most serious behaviour problem among first year pupils, and the fifth most serious problem among pupils in their fifth year of school. This echoes previous work by Kyriacou and Sutcliffe (1978a) that the highest-rated source of stress was 'pupils' poor attitudes to work'. Other studies in this area refer to teachers' concern to maintain high standards, or concern for pupils' learning (Pratt, 1976).

Teachers' concern with pupils' behaviour may also be seen to contribute towards job satisfaction. Freeman (1987) has argued that for most teachers, job satisfaction lies in the experience of teaching itself and in the 'positive feedback' that comes from a successful lesson or series of lessons. Therefore, events that may interfere with this feedback (e.g. poor attitudes and behaviour of pupils) could be a cause of job dissatisfaction. In addition, Mancini, Wuest, Clark and Ridosh, 1982; Mancini et al., 1984) have suggested that 'burnout' may be associated with the breakdown in the teacher–pupil relationship, as they found that 'burned out' teachers gave significantly less praise and information and showed less acceptance of their students.

A further problem leading to undesirable stress outcomes in teachers is the threat of actual violence. By examining medical records of teachers in the United States who had been subjected to physical or threatened assault, Bloch (1978) found that they had suffered symptoms similar to 'post-traumatic combat neurosis'. Much research has documented violence as a source of stress in school settings (e.g. ILO, 1981; Hammond, 1983; Wanberg, 1984).

A recent study of 101 secondary teachers found that half reported being bullied at least once by a pupil in the preceding term, with 10% experiencing several incidents a week (TES, 1998c).

The type of school as an influence on teacher stress

One question that needs to be addressed in the study of teacher stress is whether or not the type of school has an effect. This is because there are a number of differing characteristics (i.e. the size of the school, the pupil to teacher ratio, the age of the pupils, the academic pressures) between school types (e.g. primary and secondary schools) that may create problems for teachers, There is an assumption that certain types of school (e.g. inner-city, special education) create stress. The majority of studies have considered the effects of teaching in special education (e.g. Jones, 1971; Pratt, 1978), and the problems in dealing with pupils with learning difficulties.

Galloway et al. (1982) suggested that teacher stress might be mediated by the school organization and school climate, despite objectively unfavourable conditions. Pratt (1978) attempted a rather different approach, hypothesizing that teacher stress increased in 'poor' schools, as pupil age increased. In this small-scale study of 124 primary teachers, 'poor' schools were identified by the number of school meals given free.

Pratt (1978) found that the older the class, the greater the stress reported by the teacher, which would suggest that secondary teaching, would be the most stressful.

Much research has focused on urban schools as being at most risk (Feitler and Tokar, 1982) and recent research does appear to reinforce these findings (Leitman, Binns and Duffett, 1995; Abel and Sewell, 1999).

Teacher stress also appears to be more prevalent in larger school systems than in smaller schools (Reese and Johnson, 1988; Green-Reese, Johnson and Campbell, 1991) though there is some evidence to suggest that small rural schools need closer examination (Eastern Daily Press, 1999).

Relationships with management

Another relevant feature is that of leadership style, as this is a potential source of stress for employees in whatever type of occupation. The effects of exposure to an authoritarian style of leader have been well documented by Lewin, Lippitt and White (1939). If a head teacher, for example, does not engage in participation, encourage feedback on his or her own decisions or performance, and does not give recognition for good work, the head teacher/teacher relationship could be at risk. The reactions to this type of leadership style may vary from being passive and repressive (e.g. resulting in elevated levels of blood pressure) to anger and more overt displays of conflict (e.g. aggression). This can create stressful situations for all teachers in that school. It must be noted, however, that the actual climate or culture of the school may encourage particular forms of management style.

Career development

Teaching has always been believed to be a very secure job, and yet increasingly this is not necessarily the case (e.g. TES, 1999b). The insecurity of teachers' jobs is well documented (e.g. Needle et al., 1981; Wanberg, 1984). Individuals having to relocate are particularly vulnerable to stress as actual job change is a potential source of high stress (Lazarus, 1981). In addition, the rapid pace of change within teaching, both in terms of the nature and requirements of the job, and the technologies and materials that they have to deal with, means that teachers need to consider retraining and possible career change. There is a greater emphasis now on teacher competence and the potential for sacking and dismissal (TES, 1999b).

Status incongruence is also relevant to the section concerning relationships at work, and refers to the situation where the actual status bestowed on individuals do not match their status expectations and beliefs. This is of particular relevance to teachers at the moment, as they complain they are suffering from a poor public image in terms of prestige, salary and respect for professional status of teachers (e.g. Laughlin, 1984; Wanberg, 1984).

Under-promotion has also been found to be related to stress in teachers (Fimian, 1983; Wanberg, 1984). Thwarted ambitions are a cause of job insecurity. Criteria for promotion are unpredictable and uncertain, and this reinforces an external locus of control in the individual (Kyriacou and Sutcliffe, 1979b). Other problems may result from discrimination resulting in restricted job mobility for women (Wanberg, 1984) and lack of training for career development. Further related to this is the lack of advancement opportunities (Eskridge, 1984; Mykletun, 1984).

Organizational structure and climate

Another feature important in determining the levels of teacher stress is the structure and climate of the school in which they work. Travers and Cooper (1993) found that pressure from the structure and culture of the school was the major statistical predictor of teacher job dissatisfaction.

To some extent, cultural problems are an issue in schools. Many changes that have taken place have changed the schools' ethos (e.g. financial management, appraisal, changes in the curriculum). Many teachers complain that teaching is not what it used to be. There has also been an increase in the pastoral aspects of the job. A further problem is that new entrants to the profession will have a different expectation of what the job entails compared to the older teachers – which may create a conflicting culture within some schools.

Other potential stressors include poor communication, an inadequate amount of feedback about performance, inaccurate or ambiguous measurement criteria for performance and unfair control systems (Brief, Schueler and Van Sell, 1981). Other features that may be relevant to teachers at present are those concerning participation in decision-making, lack of effective consultation and communication, and restrictions on behaviour (e.g. lack of sanctions to deal with unruly pupils). Teachers have recently been expressing resentment at the lack of involvement in many of the changes that have taken place within education and, consequently, their schools. Traditionally the job of the teacher has been one that involved a great deal of autonomy. In the light of the changes that have taken place within education, it may be that this is yet another source of pressure for teachers.

The process of being evaluated by others can be a very stressful experience for some people, especially if the result of the evaluation has an effect on job prospects and career progression (Baron, 1986). In addition to this formal appraisal, the very job of a teacher necessitates that they are on display all of the day, in front of the pupils. Their actual performance is to a large extent evaluated every time a pupil takes an examination, or parents visit for a 'parents' evening'.

Additional pressures facing teachers are those consisting of social pressures, that is, legislation which limits responses to social situations (Needle et al., 1981), the financial and social deprivation of children (Pratt, 1978; Tellenback et al, 1983), parent–pupil relationships (Wilson, 1980), the macro-environment (Pettigrew and Wolf, 1981), and public pressures (Instructor, 1979). This suggests that teachers' problems do not just result from limitations within their own organization but also from the structure and climate of society.

The home–work interface

So far this chapter has concentrated on the sources of pressure in the teacher's working environment. There are, however, potential stressors that exist in the individual teacher's life, outside of their workplace, affecting an individual's behaviour at work, requiring consideration when assessing the sources and impact of work stress. Potential stressors include stressful life events, pressure resulting from conflict between organizational and family demands, financial difficulties, and conflicts between organizational and personal beliefs. Events occurring in the home may be both a source of stress and a source of support, just like relationships at work, and may also mitigate or exacerbate the effects of stressors experienced in the work environment.

One aspect of home life that may help exacerbate pressure is that of being part of a dual-career couple. In a profession such as teaching that has such a large proportion of women, this is bound to be a feature to be considered in the teacher stress phenomenon. One of the major problems facing dual-career families is that of society's attitude towards them (i.e. the 'traditional' family set-up is still regarded as the norm).

A study by Cooke and Rousseau (1984) revealed that although the demands of the family conflicted with the demands of work, the family also provided comfort and support. This enabled them to overcome some of the more harmful physical effects that are usually associated with stress (i.e. headaches, loss of sleep) better than teachers who were still single. There is a problem in drawing conclusions from findings such as these, in that there are probably many differences between the lifestyles of unmarried and married teachers.

Changing perceptions of the teacher

By far one of the most alarming changes for teachers has been what they see as a change in the attitude of society towards them and the job that they do. Ranjard (1984) comments that 'Teachers are persecuted by the development of a society which forces profound changes upon their profession.' Esteve (1984) points out that the traditional stereotype of the teacher as being one of friend and adviser, dedicated to helping and relating to students, maintaining an attitude of service both inside of school and out, is being replaced by a media-based stereotype, of strikes, lack of effective training, physical violence in the classroom, dismissals and poor conditions. A key additional stressor for teachers is probably that teacher training has tended to over-promote this first stereotype, whilst neglecting the second, thus not always adequately preparing new teachers for the real strains of the job.

What characteristics of the individual teacher affect how they respond to stress?

Evidence shows that in certain environments and under particular levels of pressure, some individuals survive the strain while others do not. Therefore, the experience of stress can be a very personal thing with stress resulting from an individual teacher perceiving the stressors and threats as far outweighing their available resources to meet the demands. Some teachers are therefore more susceptible than others, but we need to discover why this is the case and focus on features of the environment and the teacher that lead to reduced resistance and increased vulnerability. There are many features that predispose a teacher to deal with stress in a particular way (age, experience, life events, life stages and ability, personality, behavioural disposition, attitudes, values and needs).

Gender, cognitive style and a range of personality variables have been found to be linked to significant variations in the levels of occupational stress reported by teachers (e.g. Pierce and Molloy, 1990; Travers and Cooper, 1993; Borg and Riding, 1991) The following addresses some of these in more detail.

Age, experience and level of ability and its relationship to the experience of teacher stress

Age and experience have also been linked to the experience of stress in teaching in that it has often been suggested that the highest levels of stress might be experienced by recent entrants to the profession (usually younger teachers). This may be due to the fact that they have not yet acquired the expertise required to cope with the job. A study by Coates and Thoresen (1976) concluded that younger and less experienced teachers felt greater stress than their colleagues from pressures associated with discipline, poor promotion prospects and management issues.

Edworthy (1988) discovered that a major source of stress for younger teachers was pupils' general low ability. An Australian study by Laughlin (1984) suggested that the chief concerns of younger teachers are the pupils, whereas for those in their middle years the major source of stress is career aspects and the actual teaching itself is the problem for older teachers. A look at the literature in general highlights the issue that middle-aged teachers may face, a fear of obsolescence linked with their mid-life crisis. Therefore, as Warnat (1980) explains, this may result in middle-aged teachers worrying that their skills are somewhat outdated and what experience they have being of little value to the profession (Warnat, 1980). It is difficult when attempting to compare the experiences of younger and older teachers to eradicate the effect of actual experience and so often it is necessary to consider these two together. Authors of papers may not always make clear this point, but generally speaking this is the case in the majority of the research in this area.

The implications of the findings so far are that as teachers learn to cope with the particular stressors they face at one level, they then move on to another concern. The aforementioned study by Laughlin (1984) was not a longitudinal one and therefore the changing concerns of the same teacher throughout his or her career are not presented. Dunham (1984) has attempted to explain how apparently skilled teachers may become stressed. He explains that skilled and experienced teachers facing changes in external demands may become stressed when they discover that their previously developed coping skills are largely inadequate.

The effects of gender on the experience of teacher stress

Studies have indicated that there are gender differences in the experience of stress in teaching, one of these being with regard to job satisfaction. Researchers have reported that women teachers report greater dissatisfaction than their male colleagues with regard to classroom situations and pupil behaviour, whereas male teachers tend to report higher dissatisfaction with administration, participation and need for professional recognition and their career situation (Kyriacou and Sutcliffe, 1978b; Laughlin, 1984). Also, women tend to report higher levels of satisfaction from the job (Maxwell, 1974; Laughlin, 1984; Patton and Sutherland, 1986). However, the greatest levels of overall job satisfaction were reported by male teachers in large comprehensive schools. Female teachers in primary schools reported the least dissatisfaction (Cox, Mackay, Cox, Watts and Blockley, 1978).

One of the problems in interpreting male/female differences is that female teachers have a tendency to be primary school teachers, and there has been no direct comparison between females in primary and secondary schools. This sex bias in school type may also explain why females report a greater level of satisfaction overall than males, due to the fact that they have probably been largely drawn from the primary sector, which tends to exhibit greater levels of satisfaction anyway (Rudd and Wiseman, 1962; Maxwell, 1974; Laughlin, 1984; Patton and Sutherland, 1986). It is also evident from the literature that teachers who report high levels of satisfaction can also report high levels of stress (Kyriacou, 1987). It does not follow, therefore, that the more satisfied women may also be the least stressed.

In terms of mental well-being, studies have revealed that the reported incidence of headaches, tearfulness and exhaustion is higher among female teachers (Kyriacou and Sutcliffe, 1978b; Dunham, 1984). A point regarding the emotional expression of female teachers is that they perhaps use this as a coping mechanism, and find open expression easier than their male counterparts. A further confounding variable could be that female teachers are more able to admit to stress. In addition, research has shown that they suffer more than men from 'minor-mood' disorders and from depression, though there is not a preponderance of females among cases treated by psychiatrists (Goldberg and Huxley, 1980). Though it is important to look at male and female differences in teaching experience, care must be taken when interpreting manifestations of stress between the sexes and the controls employed. Travers and Cooper found that in their study of 1790 teachers, although the mental ill health of women was found to be higher, both male and female teachers were suffering from alarmingly high levels of mental ill health and all of its sub-scales (as measured by the Crown-Crisp Experiential Index) than all of the provided norms and other comparable occupational groups such as doctors and nurses (Travers and Cooper, 1993).

An age/sex interaction has been uncovered in relation to absence rates and decisions to leave the profession. Young female teachers have higher voluntary absenteeism than their colleagues, in particular their married female colleagues with children (Simpson, 1962, 1976). According to a DES survey (1973), young female teachers tend to report a desire to leave teaching more often than males, although the rationale behind this decision may be positive (i.e. to have a baby) rather than negative (Nias, 1985). However, in contrast to these findings, Kyriacou and Sutcliffe

(1978b) found that fewer women than men reported an intention to leave the profession, though they too found that having a baby was the most common reason for female teachers' intention to leave. This issue needs further probing in order to gain a greater insight into the motivation of female teachers, but it may well be the case that stress at work makes women choose to leave the profession rather then purely for maternity leave. Men in these studies gave as their reasons to leave, poor salary, poor promotion prospects and general dissatisfaction.

Turning to the actual sources of stress, a study of 493 Australian teachers by Laughlin (1984) found that women reported more stress concerning pupils and curriculum demands whereas men emphasized participation and professional recognition. A well-documented explanation of the differences in their concerns is that women are greatly under-represented in promoted teaching posts.

What impact does the personality and style of the teacher have?

Type A behaviour

Type A behavioural style is one of the most widely investigated 'person-based' characteristics that may influence the stress relationship (Ivancevich and Matteson, 1984).

How would we identify the type A teachers in their work setting? Type A's may be identified as those who:

1. Work long hours constantly under deadlines and conditions of overload.
2. Take work home on evenings and at weekends; they are unable to relax.
3. Often cut holidays short to get back to work, or may not even take a holiday.
4. Constantly compete with themselves and others; also drive themselves to meet high, often unrealistic standards.
5. Feel frustrated in the work situation.
6. Are irritable with work efforts and their pupils.
7. Feel misunderstood by their head teachers.

Travers and Cooper (1993) revealed that it is not always helpful to look at the concept of type A as a whole but rather the 'sub-scales' within it. They found a significantly large proportion of teachers in their sample were strong type A's. In addition, by breaking down the overall scale of type A (as designed by Bortner, 1969), greater predictability of mental ill health could be achieved. It was discovered that high levels of 'ambition and competitive behaviour' were in fact positively related to better mental health, whereas 'time consciousness/impatient behaviour' was by far the strongest predictor of mental ill health. This is helpful as it means that greater focus can be placed on these more detrimental aspects of the type A personality so that effective coping strategies can be developed.

Eysenck Personality Inventory

Two studies of particular importance in relation to teacher personality and stress are those of Pratt (1976) and Kyriacou and Sutcliffe (1979b). In the first study, Pratt utilized the Eysenck Personality Inventory (EPI) on 124 primary school teachers and reported significant correlations between reported stress and both neuroticism and extraversion. A problem does occur when interpreting these results with regard to neuroticism, as stress has been found to be positively related to neuroticism scores, as they increase when the individual is experiencing stress. This suggests that they may be measuring similar aspects of the individual (Humphrey, 1977).

Locus of control

In a study of 130 UK secondary teachers, Kyriacou and Sutcliffe (1979b) investigated the locus of control effect as revealed by a number of authors (e.g. Chan, 1977). This is where a person experiences stress related to the degree to which they perceive themselves as having a lack of control over the potentially threatening situation. It is believed that individuals with a high internal locus of control are more robust to stress than those with a high external locus of control because they tend to feel more in control of their life generally and thus better able to do something about stress when it occurs. This may be due to a better ability to deal with the demands faced or the more effective and adequate coping mechanisms being available. These researchers looked at the association between self-reported teacher stress and Rotter's Internal-External Locus of Control (I-E) scale (Rotter, 1966). A Likert scale (i.e. 1–5 from strongly agree to strongly disagree) was employed in the study and a significant correlation was obtained between self-reported teacher stress and externality. Therefore, teachers exhibiting external control reported higher levels of teacher stress. The key issue here of course is that due to so many changes in the profession even internal types may be feeling that so much is outside of their control in teaching today.

Coping methods employed by teachers

The personal coping strategies employed by teachers to deal with stress have not received a great deal of attention (Pithers and Soden, 1998b).

Studies have focused on range and intensity of stressors and degree of distress and burnout and symptoms, and though personal factors and an interactive model have been increasingly accepted in the literature there has been little in the way of research.

Recently a focus has been placed on personal resources that an individual may employ to cope with stress (Pithers and Soden, 1998b). Bowers (1995) has found that teachers who adopted an assertive/persuasive coping style when dealing with a range of stressors experienced less psychological stress. Cockburn (1996) studied 335 primary teachers and found that they were aware of 35 stress reducing strategies. 'Destressors' in the teachers' environment are collegial support and praise/recognition (Punch and Tutteman, 1996) and research evidence has revealed that social support merits further study (Pithers and Fogarty, 1995; Pithers and Soden, 1998b).

However, though it is important to consider the coping needs and styles of the individual teacher, we must remember the interactive model and not underestimate the importance of organizational stress management interventions (see Travers and Cooper, 1996).

Where should teacher stress research be heading?

This chapter has attempted to summarize the vast amount of research into the area of teacher stress – research that has been prolific over the last three to four decades. Advances in terms of methods employed have been made with research being no longer largely anecdotal in nature, but employing a variety of approaches. More regard is centred on the interactive nature of teacher stress with progression from mainly examining symptoms and causes to paying much overdue attention to the role that the individual teacher makes to the stress equation.

With many changes in education have come changes in focus of those at risk. For example, the work of Travers and Cooper (1993) following the Education Reform Act in 1988 found that head teachers were the least at risk in their sample, but time has taken its toll and the stressors they are experiencing are being studied more extensively.

However, we must not be complacent. Guglielmi and Tatrow (1999) suggest that teacher stress research has not moved on far enough and that the area is beset with methodological and conceptual difficulties. Though a link between occupational stress and burnout and ill health in this profession is suggested, they claim that research has relied on cross-sectional retrospective designs and self-report measures. They plea for greater focus on theoretical frameworks to guide research in this area –

more theory-based investigations that test causal models of teacher stress and health with more sophisticated research designs and measurement strategies.

To some extent these comments are fair. There still needs to be more longitudinal research that examines the process of teachers becoming stressed – there are so many factors that the picture does tend to be a rather cluttered one. Advances in psychophysiological measures of stress still need to be refined. Much work is still based largely on self-report data and, though we should assume that teachers should be the best people to report on their own experiences, a concern must be expressed here. The vast amount of research into this area coupled with the media focus has to a large extent created a 'negative' industry – we now assume that teachers are stressed and researchers and teachers alike tend to start from this premise.

Cross-cultural research is on the increase and we need to do more of it, not just because it makes for interesting (and publishable) papers but because we could perhaps learn about improving the teacher's working life from other countries who are experiencing problems.

Though more attention is paid to it, there is still much work to be done in exploring the role of personality and individual coping styles. Teachers need not feel that this will be admitting blame; it is necessary so that more effective targeted stress management techniques can be recommended. In parallel with this, much work is being done by Local Education Authorities and counselling and support-lines across the country and large amounts of data are being gathered. Perhaps it is time that academics and practitioners worked more closely in collaboration to ensure well-designed studies, which are to a large extent theory driven, but highly pragmatic in terms of outcomes. There is also a need for more research to evaluate stress management interventions with teachers.

It is important to recognize that in all of this we are dealing with a real stress problem that has major ramifications for teachers, pupils, schools and our education provision as a whole. More research is needed but we do also need to be tackling the problem of how to improve the situation with some urgency. How much more time can we spend on tightening up our research when the problem of teacher stress is getting more prevalent. As researchers in the field we need to ask ourselves what positive impact have we had on the situation over these last thirty or so years? Have we exacerbated it?

The overall biggest concern must be that as concepts 'teaching' and 'stress' are now almost synonymous. This could actually be creating an unhelpful focus as much energy is devoted to reiterating what actually causes stress instead of dealing with more constructive aspects, e.g. problem-solving, prioritizing, self-management and skills training.

An area where research should be heading is in the area of the 'healthy school'. More and more evidence is suggesting that schools and the associated pressures are not always psychologically healthy places for staff and pupils (Hart, 1994). We need to be investigating the actual cultures of schools for all of their members at all levels and try to find ways of creating healthier school environments. A rather daunting but challenging and necessary task.

References

Abel MH, Sewell J (1999) Stress and burnout in rural and urban secondary school teachers. Journal of Educational Research 92(5): 287.

Austin DA (1981) The teacher burnout issue. Journal of Physical Education, Recreation and Dance 52(9): 35–56.

Bacharach SB, Bauer SC, Conley S (1986) Organisational analysis of stress: the case of elementary and secondary schools. Work and Occupations 13(1): 7–32.

Baron RA (1986) Behaviour in Organisations, 2nd edn. Newton, MA: Allyn & Bacon.

Bayer E, Chauvet N (1980) Libertés et constraintes de l'exercise pédagogique, Faculté de Psychologie et Sciences de l'Education, Genève.

Begley D (1982) Burnout among special education administrators. Paper presented at the Summer Convention of the Council for Exceptional Children, Houston, Texas

Blase JJ (1982) A social-psychological grounded theory of teacher stress and burnout. Educational Administration Quarterly 18(4): 93–113.

Bloch A (1978) Combat neurosis in inter-city schools. American Journal of Psychiatry 135: 1189–92.

Borg MG, Riding RJ (1991) Towards a model for the determinants of occupational stress among school teachers. European Journal of Psychology of Education 6: 355–73.

Borg MG, Riding RJ, Falzon JM (1991) Stress in teaching: A study of occupational stress and its determinants, job satisfaction and career commitment among primary schoolteachers. Educational Psychology: An International Journal of Experimental Educational Psychology II: 59–75.

Bortner RW (1969) A short rating scale as a potential measure of pattern a behaviour. Journal of Chronic Diseases 22: 87–91.

Bowers T (1995) Teacher stress and assertiveness as a coping mechanism. Research in Education 53: 24–30.

Brenner SD, Sorbom D, Wallius E (1985) The stress chain: A longitudinal study of teacher stress, coping and social support. Journal of Occupational Psychology 58: 1–14.

Bridges EM (1980) Job Satisfaction and teacher absenteeism. Education Administration Quarterly 16: 41–6.

Brief AP, Schuler RS, Van Sell M (1981) Managing Job Stress. Boston, MA: Little Brown.

Brown M, Ralph S (1997) Change-linked work-related stress in British schools, British Psychological Society (Ed Section) Annual conference, 14–16 November.

Burke E, Dunham J (1982) Identifying stress in language teaching. British Journal of Language Teaching 20: 149–52.

Burke RJ, Shearer J, Deszca G (1984) Correlates of burnout phases among police officers. Group and Organisational Studies 9: 451–66.

Cains RA, Brown CR (1998) Newly qualified teachers: A comparison of perceptions held by primary and secondary teachers of their training routes and their early experiences in post. Educational Psychology 18(3): 341–52.

Chan DW (1998) Stress, coping strategies and psychological distress among secondary school teachers in Hong Kong. American Educational Research Journal 35(1): 145–63.

Chan KB (1977) Individual differences in reactions to stress and their personality and situational determinants: some implications for community mental health. Social Science and Medicine 11: 89–103.

Claggett CA (1980) Teacher stress at a community college: Professional burnout in a bureaucratic setting, Education Resources Information Centre, No. 195310.

Clark EH (1980) An analysis of occupational stress factors as perceived by public school teachers. Unpublished Doctoral Dissertation, Auburn, AL: Auburn University.

Claxton G (1988) The Less Stress Workshop. Copyright G. Claxton, Kings College, London.

Coates TJ, Thoresen CE (1976) Teacher anxiety: a review with recommendations. Review of Educational Research 46(2): 159–84.

Cockburn AD (1996) Primary teachers' knowledge and acquisition of stress relieving strategies. British Journal of Educational Psychology 66(3): 399-410.

Comber L, Whitfield R (1979) Action on Indiscipline: A Practical Guide for Teachers. Hemel Hempstead: NASUWT.

Connors SA (1983) The school environment: A link to understanding stress. Theory in Practice 22(1): 1–20.

Cooke RA, Rousseau DM (1984) Stress and strain from family roles and work role expectations. Journal of Applied Psychology 69: 252–60.

Cooper CL, Kelly M (1993) Occupational stress in headteachers. British Journal of Educational Psychology 63: 130–43.

Cox T (1977) The nature and management of stress in schools in Clwyd County Council. In: The Management of Stress in Schools. Conference report prepared by Clwyd County Council Department of Education.

Cox T, Mackay CJ, Cox S, Watts C, Brockley T (1978) Stress and well-being in school teachers. Paper presented to the Ergonomics Society Conference on Psychophysiological Response to Occupational Stress, Nottingham University, Nottingham, England. (Cited in Litt MD, Turk DC (1985) Sources of stress and dissatisfaction in experienced high school teachers. Journal of Educational Research 78(3): 178–85.)

Crane SJ, Iwanicki EF (1986) Perceived role conflict, role ambiguity and burnout among special education teachers. Remedial and Special Education (RASE) 7(2): 24–31.

Crown S, Crisp AH (1979) Manual of the Crown-Crisp Experiential Index. London: Hodder & Stoughton.

D'Arienzo RV, Moracco JC, Krajewski RJ (1982) Stress in Teaching. Washington, DC: University Press of America.

DES (1973) Teacher Turnover Reports in Education, No 79. London: Department of Education and Science.
Dunham J (1976) Stress situations and responses. In: NAS/UWT, Stress in Schools. Hemel Hempstead: NAS/UWT.
Dunham J (1977) The signs, causes and reduction of stress in teachers. In: The Management of Stress in School. Mold: Clwyd County Council
Dunham J (1980) An exploratory comparative study of staff stress in English and German comprehensive schools. Educational Review 32: 11–20.
Dunham J (1983) Coping with Stress in Schools. Special Education: Forward Trends 10(2): 6–9.
Dunham J (1984) Stress in Teaching. Beckenham: Croom Helm.
Dunham J (1992) Stress in Teaching, 2nd edn. London: Routledge.
Eastern Daily Press (1999) Breaking point: Crisis in small schools. 8 November, p. 1.
Education Reform Act (1988) London: HMSO.
Edworthy A (1988) Teaching can damage your health [feature on a research report]. Education 8 January.
Eskridge DH (1984) Variables of teacher stress symptoms, causes and stress management techniques. Unpublished Research Study, Commerce, TX: East Texas State University.
Esteve JM (1989) Teacher burnout and teacher stress. In: Cole M, Walker S (eds) Teaching and Stress. Milton Keynes: Open University Press.
Esteve JM (1984) L'image des enseignants dans les moyen de communication de masse. European Journal of Teacher Education. 7(2): 203–9.
Feitler FC, Tokar EB (1982) Getting a handle on teacher stress: How bad is the problem? Educational Leadership 39: 456–8.
Fielding JE (1982) Personality and situational correlates of teacher stress and burnout. Doctoral Dissertation, Eugene, OR: University of Oregon; Dissertation Abstracts International, 43/02A.
Fimian MJ (1983) A comparison of occupational stress correlates as reported by teachers of mentally retarded and non-mentally retarded handicapped students. Education and Training of the Mentally Retarded 18(1): 62–8.
Fimian MJ, Santoro TM (1983) Sources and manifestations of occupational stress as reported by full time special education teachers. Exceptional Children 49(6): 540–3.
Freeman A (1987) Pastoral care and teacher stress. Pastoral Care in Education 5(1): 22–8.
Galloway D, Panckhurst F, Boswell K, Boswell C, Green K (1982) Sources of Stress for Class Teachers. National Education 64: 166–9.
Goble NM, Porter JF (1980) La Cambiante Function de Profesor. Madrid: Narcea.
Goldberg D, Huxley P (1980) Mental Illness in the Community: The Pathway to Psychiatric Care. London: Tavistock Publications
Golembiewski RT, Munzenrider R, Carter D (1983) Phases of progressive burnout and their worksite covariates. Journal of Applied Behavioural Sciences 19: 461–81.
Green-Reese S, Johnson DJ, Campbell WA (1991) Teacher job satisfaction and teacher job stress: School size, age and teaching experience. Education 112(2): 247–52.
Guglielmi RS, Tatrow K (1999) Occupational stress, burnout and health in teachers: a methodological and theoretical analysis. Review of Educational Research 68: 61–9.

Hammond JM (1983) School improvement using a trainer of trainers approach: reducing teacher stress. Journal of Staff Development. 4(1): 95–100.

Hargreaves D (1978) What teaching does to teachers. New Society 9(43): 540–3.

Hart PM (1994) Teacher quality of work life: Integrating work experiences, psychological distress and morale. Journal of Occupational and Organisational Psychology 67: 109–32.

Hawkes RR, Dedrick CV (1983) Teacher stress: phase II of a descriptive study. National Association of Secondary School Principals Bulletin 67(461): 78–83.

Hembling DW, Gilliland B (1981) Is there an identifiable stress cycle in the school year. Alberta Journal of Educational Research 27(4): 324–30.

Humphrey M (1977) Review – Eysenck Personality Questionnaire. Journal of Medical Psychology 50: 203–4.

ILO (1981) Employment and conditions of work of teachers. Geneva: International Labour Organisation.

Instructor (1979) Teacher burnout: How to cope when the world goes black, Instructor 88(6): 56–62.

Ivancevich JM, Matteson MT (1984) A type A-B person-work environment interaction model for examining occupational stress and consequences. Human Relations 37(7): 491–513.

Kalker P (1984) Teacher stress and burnout: Causes and coping strategies. Contemporary Education 56(1): 16–19.

Kelly AV (1974) Teaching Mixed Ability Classes. London: Harper & Row.

Kyriacou C (1981) Social support and occupational stress among schoolteachers. Educational Studies 7(1): 55–60.

Kyriacou C (1987) Teacher stress and burnout: an international review. Educational Research 29(2): 146–52.

Kyriacou C (1997) Effective Teaching in Schools, 2nd edn. Cheltenham: Stanley Thorne.

Kyriacou C (1998) Teacher stress: past and present. In: Dunham J, Varma V (eds) Stress in Teachers: Past, Present and Future, pp. 1–13. London: Whurr Publishers.

Kyriacou C, Roe H (1988) Teachers' perceptions of pupils' behaviour problems at a comprehensive school. British Educational Research Journal 14(2): 167–73.

Kyriacou C, Sutcliffe J (1978a) Teacher stress: prevalence, sources and symptoms. British Journal of Educational Psychology 48: 159–67.

Kyriacou C, Sutcliffe J (1978b) A model of teacher stress. Educational Studies 4: 1–6.

Kyriacou C, Sutcliffe J (1979a) Teacher stress and satisfaction. Educational Research 21(2): 89–96.

Kyriacou C, Sutcliffe J (1979b) A note on teacher stress and locus of control. Journal of Occupational Psychology 52: 227–8.

Laughlin A (1984) Teacher stress in an Australian setting: the role of biographical mediators. Educational Studies 10(1): 7–22.

Lawrenson GM, McKinnon AJ (1982) A survey of classroom teachers of the emotionally disturbed: attrition and burnout factors. Behavioural Disorders 8: 41–8.

Lazarus RS (1966) Psychological Stress and the Coping Process. New York: McGraw Hill.

Lazarus RS (1981) Little hassles can be hazardous to health. Psychology Today July: 58–62.

Lazarus RS, Folkman S (1984) Stress, Appraisal, and Coping. New York: Springer.

Leitman R, Binns K, Duffett A (1995) The American teacher, 1984-1995, Metropolitan Life Survey, old problems, new challenges (Report No. SP 036 532). New York: Metropolitan Life Insurance Co. (ERIC Document Reproduction Service No. ED 392-783).

Lewin K, Lippitt R, White RK (1939) Patterns of aggressive behaviour in experimentally created social climates. Journal of Social Psychology 10: 271–99.

Litt MD, Turk DC (1985) Sources of stress and dissatisfaction in experienced high school teachers. Journal of Educational Research 78(3): 178–85.

Lowenstein LF (1991) Teacher stress leading to burnout: Its prevention and cure. Education Today 41(2): 12–16.

Mancini V, Wuest D, Clark E, Ridosh N (1982) A comparison of the interaction patterns and academic learning time of low-burnout and high-burnout physical educators. Paper presented at Big Ten Symposium on Research on Teaching, Lafayette, IN.

Mancini V, Wuest D, Vantine K, Clark E (1984) Use of instruction and supervision in interaction analysis on burned out teachers: its effects on teaching behaviours, level of burnout and academic learning time. Journal of Teachers in Physical Education 3(2): 29–46.

Maxwell M (1974) Stress in schools. Centrepoint 7: 6–7.

Milaat WA (1997) Stress in schools; prevalence of hidden psychiatric illness among Jeddah school workers. Saudi Medical Journal 18(3): 240–3.

Mobley WH (1982) Employee Turnover: Causes, Consequences and Control. Reading, MA: Addison-Wesley.

Moracco JC, Gray P, D'Arienzo RV (1981) Stress in teaching: A comparison of perceived stress between special education and regular teachers. Education Information Resource Center, Alabama, USA.

Mykletun RJ (1984) Teacher stress: perceived and objective sources and quality of life. Scandinavian Journal of Educational Research 28(1): 17–45.

Nagy S (1982) The relationship of Type A personalities, workaholism, perception of the school climate and years of teaching experience to burnout of elementary and junior high school teachers in Northwest Oregon school district. Unpublished Doctoral Dissertation, University of Oregon, Eugene, Oregon.

Needle RH, Griffin T, Svendsen R (1981) Occupational stress: Coping and health problems of teachers. Journal of Health 51(3): 175–81.

Needle RH, Griffin T, Svendsen R, Berney C (1980) Teacher stress: sources and consequences. Journal of School Health 50(2): 96–9.

Nias J (1985) A more distant drummer: teacher development as the development of self. In: Barton L, Walker J (eds) Education and Social Change. Beckenham: Croom Helm.

NYSUT (1980) Disruptive students cause stress. Information Bulletin, New York State United Teachers.

Ostell A, Oakland S (1999) Absolutist thinking and health. British Journal of Medical Psychology 72: 239–50.

Patton J, Sutherland D (1986) Survey on Symptoms of Stress Among EIS Members in Clackmannanshire Schools. Educational Institute of Scotland, Clackmannanshire Branch.

Phillips BL, Lee M (1980) The changing role of the American teacher: current and future sources of stress. In: Cooper CL, Marshall J (eds) White Collar and Professional Stress. Chichester: Wiley.

Pierce CMB, Molloy GN (1990) Relations between school type, occupational stress, role perceptions and social support. Australian Journal of Education 34: 330–8.

Pithers RT, Fogarty GJ (1995) Occupational stress among vocational teachers. British Journal of Educational Psychology 65: 3–14.

Pithers B, Soden R (1998a) Scottish and Australian teacher stress and strain: a comparative study. British Journal of Educational Psychology 68(2): 269–79.

Pithers B, Soden R (1998b) Personal resources strength and teacher strain. Research in Education 60: 1.

Pratt J (1976) Perceived stress among teachers: an examination of some individual and environmental factors and their relationship to reported stress. Unpublished MA thesis, University of Sheffield.

Pratt J (1978) Perceived stress among teachers: the effects of age and background of children taught. Educational Review 30: 3–14.

Punch KF, Tutteman E (1996) Reducing teacher stress: the effects of support in the work environment. Research in Education 56 63–72.

Ranjard P (1984) Les enseignants persecutes. Paris: Robert Jauze.

Reese SA, Johnson DJ (1988) School size and teacher job satisfaction of urban secondary school physical education teachers. Education 108: 382–4.

Rotter JB (1966) Generalised Expectancies for Internal Versus External Control of Reinforcement. Psychological Monographs 80 1 (whole number 609) 1–28.

Rudd WD, Wiseman S (1962) Sources of dissatisfaction among a group of teachers. British Journal of Educational Psychology 32(3): 275–91.

Russell DW, Altmaier E, Van Velzen D (1987) Job-related stress, social support and burnout among classroom teachers. Journal of Applied Psychology 72(2): 269–74.

Schwab RL (1981) The relationship of role conflict, role ambiguity, teacher background variables and perceived burnout among teachers, Doctoral Dissertation, Storrs, CT: University of Connecticut; Dissertation Abstracts International, 41 (09-A), (2) 3823-a.

Schwab RL (1983) Teacher burnout: Moving beyond 'psychobabble'. Theory into Practice 22: 21–5.

Schwab RL, Iwanicki EF (1982) Perceived role conflict, role ambiguity, and teacher burnout. Education Administrative Quarterly 18: 60–74.

Simpson J (1962) Sickness absence in teachers. British Journal of Industrial Medicine 19: 110–15.

Simpson J (1976) Stress: Sickness Absence in Teachers. In: The Management of Stress in Schools. Hemel Hempstead: NAS/UWT.

Smith J, Cline D (1980) Quality program. Pointer 24(2): 80–7.

Spanoil L, Caputo GG (1979) Professional Burnout: A Personal Survival Kit. Lexington, MA: Human Services Associates.

Sparks DC (1979) A biased look at teacher job satisfaction. Clearing House 52(9): 447–9.

Sparks DC, Hammond J (1981) Managing teacher stress and burnout, Washington, DC: Educational Information Research Center.

Spuck DW (1977) Rewards structure in the public high school. Educational Administration Quarterly 18–34.

Steers R, Mowday R (1981) Employee Turnover and post decision accommodation process. In: Staw B, Cummings I (eds) Research in Organisational Behaviour, vol. 3. Greenwich: JAI Press.

Tellenbeck S, Brenner SO, Lofgren H (1983) Teacher stress: exploratory model building. Journal of Occupational Psychology 56: 19–33.

TES (1994) Extra pressure piled for Catholic heads. Times Education Supplement 18 November.

TES (1997a) Ill-health blamed for exodus of staff. Times Education Supplement 12 September.

TES (1997b) Concern as stress and depression set in early. Times Education Supplement 3 October.

TES (1997c) Stressed-out heads reach for Prozac. Times Education Supplement 7 November.

TES (1998a) Inspections clog your arteries. Times Education Supplement 27 February.

TES (1998b) What gives you job satisfaction? Times Education Supplement 12 June.

TES (1998c) Grown-up victims of classroom bullying. Times Education Supplement 26 June.

TES (1998d) Blunkett bids to purge the sick. Times Education Supplement 31 July.

TES (1999a) Staff under stress to stub out fags. Times Education Supplement 9 July.

TES (1999b) Schools get tough with bad teachers. Times Education Supplement 10 September.

Travers CJ (1997) Vive la difference. Paper presented to the British Psychological Society, Annual Occupational Psychology Conference, Blackpool, January 1997.

Travers CJ, Cooper CL (1993) Mental ill health, job satisfaction, alcohol consumption and intention to leave in the teaching profession. Work and Stress 7(3): 203–19.

Travers CJ, Cooper CL (1994) Psychophysiological responses to teacher stress: A move towards more objective methodologies. European Review of Applied Psychology 44(2): 137–46.

Travers CJ, Cooper CL (1996) Teachers under Pressure: Stress in the Teaching Profession. London: Routledge.

Usher J (1998) Workplace stress and the law. In: Dunham J, Varma V (eds) Stress in Teachers: Past, Present and Future. London: Whurr Publishers.

Wanberg EG (1984) The complex issue of teacher stress and job dissatisfaction. Contemporary Education 56(1): 11–15.

Warnat WI (1980) Teacher stress in the middle years: crises vs change. Pointer 24(2): 4–11.

Weiskopf PE (1980) Burnout among teachers of exceptional children. Exceptional Children 47: 18–23.

Westerhouse MA (1979) The effects of tenure, role conflict, and role conflict resolution on the work orientation and burnout of teachers. Doctoral Dissertation, Berkeley, CA: University of California, Dissertation Abstracts International, 41 (01A) 8014928, 174.

Wilson CF (1980) Stress profile for teachers: test manual and preliminary data. Department of Education, San Diego County, San Diego, CA.

Zabel R, Zabel MK (1982) Factors in burnout among teachers of exceptional children. Exceptional Children 49: 261–3.

Chapter 9
Organizational Work Stress Interventions in a Theoretical, Methodological and Practical Context

MICHIEL A.J. KOMPIER AND TAGE S. KRISTENSEN

Introduction

This book is on the 'past, present and future' of stress in occupations. 'Past, present and future' is also the title of a sixties hit song by the Shangrilas. Inspired by this record, in this chapter we will discuss the past, present and future of research into work stress interventions. Although we will also address person-oriented measures, the emphasis will be on the organizational level, that is work directed interventions in a theoretical, methodological and practical context.

Past

The first line of the Shangrilas' song is 'Let me tell you of the past'. What does the past of stress interventions look like? The first comprehensive and critical review of both personal and organizational strategies for handling job stress was published in 1979 by Newman and Beehr. The authors provide a general matrix for the review of adaptive responses to job stress, and then review three types of strategies: (a) personal strategies for handling job stress, (b) organizational strategies, and (c) strategies used by persons and organizations outside the focal organization to help those inside the organization manage stress. Their main conclusion is that many strategies for managing job stress exist but that: 'there is a definite lack of evaluative research in this domain. Very few of the purported strategies for handling job stress have been evaluated with any sort of scientific rigor. This is not to say that the majority of the strategies is not valid. It is to say, simply, that the evaluative research has not been done – at least in the job context' (Newman and Beehr, 1979, pp. 3–4).

   Until then, work and organizational psychologists had not been very active in stress prevention research. Newman and Beehr underline the need for industrial and organizational psychologists to get involved in this field: 'The raison d'etre for this paper is to stimulate our industrial/organizational psychologist colleagues to become involved in and to become leaders of this most important scientific endeavor' (p. 2).

   And they are rather optimistic: 'We have some good ideas of what variables are important and some good ideas about how they may be related. Gathering empirical evidence is difficult but some systematic progress is being made in some quarters. There is a considerable body of knowledge (psychological and medical) regarding how the human (mind-body) operates. It seems that all that is lacking is the hard work required to put all this together in a concerted effort to develop and evaluate the effectiveness of personal and organizational strategies for handling job stress. We are confident that I/O psychologists are up to the challenge' (Newman and Beehr, 1979, pp. 40–1).

Present

A framework: two models

Twenty years have passed since the Newman and Beehr review. Has work and organizational psychology taken up the gauntlet? We will try to answer this question, and thereby discuss the present status with respect to research into work stress interventions against the background of a general work stress model as depicted in Figure 9.1.

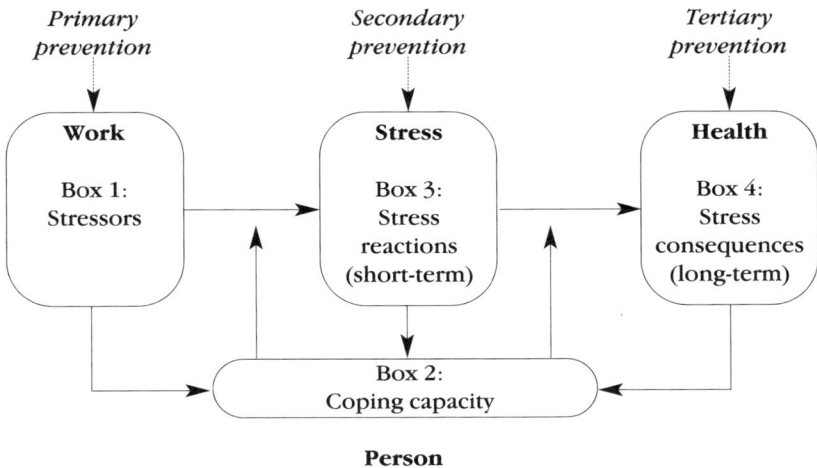

Figure 9.1: A model of work, stress and health.

This model is both simple and comprehensive. First, it shows how (i.e. as a function of both the work situation and personal characteristics of the employee) stress reactions may develop and eventually – often in a process of continual exposure and insufficient recovery – may result in long term stress consequences. Second, it provides a clear framework for analysing work stress, the two questions being 'What's in every box?' and 'How are these boxes interrelated?' (Levi, 1984).

Third, and most importantly in the context of this chapter, it provides a framework for intervention and prevention, introducing two crucial distinctions. In the first place, interventions may be directed at either the work situation (changing the work situation, box 1) or the coping capacity of the employee (changing the employee, box 2). In the second place, interventions may be aimed at (a) eliminating, reducing or altering stressors in the working situation (primary prevention), or (b) preventing employees who are already showing signs of stress from getting sick and to increase their coping capacity (secondary prevention), or (c) treating those employees who show serious stress consequences and rehabilitation after sickness absenteeism (tertiary prevention). The three vertical arrows in Figure 9.1 correspond with these three types of prevention.

By combining these two main axes, 'changing work versus changing the person' and 'eliminating risks versus preventing reactions from becoming worse', a conceptual framework can be developed which indicates four types of prevention and intervention (Figure 9.2).

Prevention

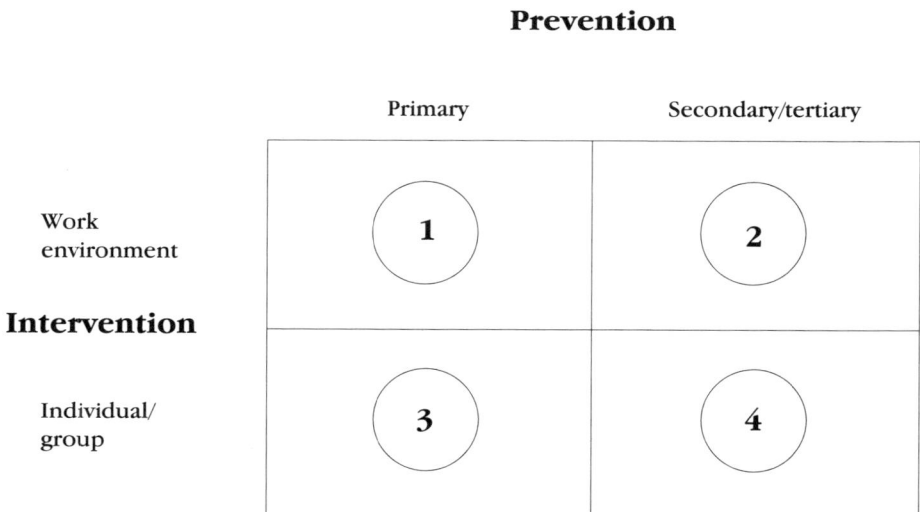

Figure 9.2: A framework for stress prevention and intervention.

   Examples of measures in the first quadrant are changing the job content (job redesign), ergonomic improvements, the introduction of autonomous teams, and new working and resting arrangements. Other examples are career development activities and increasing social support within the organization. Examples of the second quadrant are comparable measures directed at those employees who already show signs of stress, for instance special work schedules for older employees or workers recovering from a long-term sickness spell. Examples of the third quadrant are selection, pre-employment medical examination, health promotion and wellness programmes, and training programmes. Examples of the fourth quadrant are rehabilitation after sick-leave, post-traumatic stress assistance programmes, response- or symptom-directed techniques such as relaxation, and also psychotherapy. When compared to the more 'collective' measures from quadrant three, quadrant four measures are more 'tailor made'.
   With the help of Figure 9.2, it is possible to search systematically for effective strategies against work stress. The model facilitates the systematic consideration of possible changes in the work situation and with the worker, and 'forces' us to consider both 'the healthy' and 'the sick'. We should keep in mind that the same type of intervention may fit into more than one quadrant, since sometimes a given intervention (e.g. the intro-duction of autonomous teams) may be seen upon as primary prevention for one person (who is healthy) and as secondary for another person (who has stress-related health complaints).

Work stress prevention: review of the field

Since the pioneering work of Newman and Beehr (1979) both theorists and practitioners have been very active in this field and various reviews on stress prevention have been published (e.g. DeFrank and Cooper, 1987; Cooper and Payne, 1988; Ivancevich, Matteson, Freedman and Phillips, 1990; Kahn and Byosiere, 1992; Burke, 1993; Cox, 1993; Cooper and Cartwright, 1994; Reynolds and Briner, 1994; Murphy, 1984, 1986, 1996; Van der Hek and Plomp, 1997; Quick, Quick, Nelson and Hurrell, 1997; Briner and Reynolds, 1999). In overviewing the last two decades, five interrelated conclusions may be drawn.

There is a great deal of activity

Organizational stress is a rapidly expanding field and so is occupational stress management. Scientific progress may be illustrated by the publication of important standard books on theories of occupational stress (e.g. Cooper and Payne, 1988; Karasek and Theorell, 1990; Cooper, 1998), on research methodology (e.g. Kasl and Cooper, 1987; Frese and Zapf, 1988) and on preventive stress management in organizations (e.g. Quick et al., 1997). Also several new academic journals have been launched, e.g. *Work and Stress*, the *International Journal of Stress Management*, and the *Journal of Occupational Health Psychology*, which is also the label for a new interdisciplinary academic sub-discipline. In the mission statement of the latter journal 'Special attention is given to articles with a prevention basis'. Also *Work and Stress* provides a forum for both academic papers and 'scholarly articles of concern to policy-makers, managers and trades unionists who have to deal with such issues'. Another example of the increase of academic, but also practitioner endeavour in this field are the four successful APA/NIOSH Conferences on Work, Stress and Health that have been organized during the 1990s. On these conferences, an increased emphasis on prevention and intervention has been witnessed (see Quick, Murphy and Hurrell, 1992; Murphy, Hurrell, Sauter and Keita, 1995).

During the last decades, rather independently of this scientific progress, stress management has become a booming and commercial market with a 'surge of practitioner activity' (Kahn and Byosiere, 1992, p. 623). In addition to the academic and practitioner effort in stress management, occupational stress has also become a popular topic in the media. Media reports on occupational stress and its effects are numerous, not only in the US and Europe but also, for instance, in Australia and Japan. Finally, several countries, especially in the European Union, have introduced specific legal regulations with respect to the assessment and prevention of occupational stress (Kompier, De Gier, Smulders and Draaisma, 1994).

Stress reduction is primarily a 'band aid approach'

Although there is a great deal of both academic and practitioner activity, 'This activity is concentrated disproportionally on reducing the effects, rather than reducing the presence of stressors at work' (Kahn and Byosiere, 1992, p. 623). In terms of the model depicted in Figure 9.2, current stress interventions mainly constitute secondary and tertiary prevention, i.e. they are of a 'post hoc' (reactive) nature, e.g. counselling of 'stressed' employees, individual psychotherapy, relaxation, or biofeedback. Primary prevention ('rooting out the risks'), and especially measures in the first quadrant in Figure 9.2, are scarce.

Main target is the individual

Related to this, the main target is the individual employee rather than the workplace or the organization. A worker-oriented approach (quadrants 3 and 4 in Figure 9.2), for instance, by improving employees' skills to manage, resist or reduce stress ('the stress inoculation approach') is followed more frequently than a work-oriented approach, for example by changing the management style or increasing control (quadrants 1 and 2). Most programmes aim at the reduction of the cognitive appraisal of stressors and their subsequent effects (e.g. mood states, psychosomatic complaints) rather than at the reduction or elimination of the stressors themselves. Typically, these interventions are 'prescriptive, person-oriented, relaxation-based techniques, such as progressive muscle relaxation, biofeedback, meditation, and cognitive-behavioral skills training' (Murphy, 1996, pp. 113–14), or combinations of these techniques. Most of these techniques stem from the field of clinical and counselling psychology.

Lack of evaluation research, and in case of evaluations, lack of strong designs

'The surge of practitioner activity in the domain of stress management has not been accompanied by a commensurate increase in serious research; most programs have not been evaluated in that sense' (Kahn and Byosiere, 1992, p. 623). There seem to be two problems: The first is that many interventions are not evaluated in a systematic way. As Van der Hek and Plomp (1997, p. 135) have commented: 'The low proportion of [organizational level] studies including interventions directed towards these causes does not necessarily give a correct representation of their actual occurrence; a methodologically sound evaluation on organizational level is usually much more difficult to implement.' The second is that in case of evaluations, study designs are often characterized by serious methodological flaws. As to the latter, there is a considerable divergence between theoretical preaching and practice. Theory (e.g. Beehr and O'Hara, 1987; Kasl, 1987; Frese and Zapf, 1988; Spector, 1992) preaches longitudinal designs, the use of (randomized) control group(s), the use of both objective and subjective measures for identifying stressors as well as the short- and long-term consequences of stress, and the inclusion of mediating or moderating variables (e.g. need for control, negative affectivity). Practice is quite different, especially the practice of organizational level interventions (type 1 and type 2 in Figure 9.2). Let us first consider the individual-oriented approaches (types 3 and 4 in Figure 9.2). From a review by Murphy (1996) it follows that with respect to these individual-oriented techniques designed to help employees modify their appraisal of stressful situations or deal more effectively with the symptoms of stress (type 4 in Figure 9.2), gradually the picture is becoming clearer. Murphy demonstrates that there is not only an increase of published studies of this type, but also that the quality of these studies has improved with regard to methodological rigour. We should, however, bear in mind that these studies still reflect only a small proportion of the practitioner activity in this field.

Publications on quadrant 1 and quadrant 2 programmes, those that target at reducing or eliminating organizational stressors, are still rare. For example, in a 1987–1994 review by Van der Hek and Plomp (1997) – an update of the 1987 review by DeFrank and Cooper – only two of the 24 intervention studies included interventions of this type.

'One size fits all' – Systematic risk assessment is often lacking

Related to the above mentioned characteristics, there is another peculiarity with the majority of stress management programmes. They are of the 'one size fits all' and 'man for all seasons' type. As summarized concisely by Kahn and Byosiere (1992): 'Programs in stress management that are sold to companies show a suspicious pattern of variance; they differ more by practitioner than by company. When practitioners in any field offer sovereign remedies regardless of the presenting symptoms, patients should be wary' (p. 623). This 'one size (or one pill) fits all' approach to the practice of stress prevention stands in the way of a systematic risk assessment ('stress audit', stress analysis) identifying risk factors and risk groups. Such a systematic analysis is often lacking in stress intervention studies.

Current state of the art: critical evaluation

First, we will compare present with past. Next we will make a more critical evaluation of the present 'state of the art' with respect to stress prevention and its research.

During the last two decades we have definitely witnessed more theoretical and applied attention for work stress and its prevention, more empirical intervention studies and more evaluation research than earlier. These are positive developments. We also noticed that stress management has become a commercial market in which quality is at least doubtful. Although the number of published evaluation studies is increasing, they still constitute a small proportion of the current activity in stress prevention. This commercial boom without a theoretical and evidence-based foundation is a negative development. Furthermore, although there is more theoretically and methodologically sound intervention research than twenty years ago, this type of data is still piecemeal and only covers a minority of current stress prevention activities. Finally, this kind of evaluation research still has 'a post hoc bias to the individual'. Studies into work and organization directed interventions are still too scarce.

Therefore, the overall situation with respect to stress prevention is not satisfactory. We cannot yet provide clear answers to simple questions such as: Does work stress prevention work? Which program types and components are effective, and which are not? Why do certain components work? And, more specifically, how, i.e., which mechanisms are involved? Which are intended and unintended side-effects? What are costs, benefits and limitations? What are stimulating and obstructing factors? The present situation is well summarized by Griffiths and her colleagues (1996): 'The lack of evaluation of such interventions is a major problem and a significant barrier to progress in reducing work-related stress' (p. 66). It is interesting to note that this conclusion fits in the general state of the art with respect to occupational health and safety intervention studies as reviewed by Goldenhar and Schulte (1994). Goldenhar and Schulte (1994, p. 764) argued: 'The lack of sound evidence about the outcomes of these intervention studies should be viewed not as a negative assessment of their potential but rather as an opportunity to develop a new, research-based body of literature to which both academics and practitioners can continue to add new evidence concerning various aspects of occupational health and safety.'

It is an intriguing question as to why organizational level intervention studies are so scarce, and why it might be that companies express a preference for 'post hoc' individual directed interventions. At least five factors seem to contribute to this past and present preoccupation with person-oriented interventions.

The attitudes and values of company management

The content of most intervention programmes, with their emphasis on individual approaches, reflects the attitudes and values of company management. These values are rather individualistic, as are those of most higher-ranked white-collar employees. Organizational decision-makers have a tendency to explain (in more psychological terms 'to attribute') organizational successes and failures through individual characteristics of the persons involved. For instance, managers are often inclined to blame personality factors and lifestyles of employees who are absent from work or report health complaints. Less frequently the question is raised whether and how their own leadership style might affect absenteeism and health complaints. Also, decision-makers often point to the potential role of stressful life events (e.g. family problems, such as a divorce or the loss of a beloved) or responsibilities and demands in family life (e.g. raising children or caring for elderly parents). Of course, on the micro-level (i.e. on the level of the individual employee), stressors at work are often accompanied by stressors in one's family situation (e.g. Frone, Russell and Cooper, 1992), but because of the mutual influence and spillover between both domains, the causes and consequences can scarcely be disentangled. Furthermore, if one holds individual characteristics responsible for differences in experienced stress, it is difficult to explain why some occupations show significantly more stress complaints and higher sickness absence rates than others (Kompier, Geurts, Gründemann, Vink and Smulders, 1998). A serious risk attached to this view is that the employee is regarded as being 'guilty' of his or her own health problems (referred to as blaming the victim; McLeroy, Bibeau, Steckler and Glanz, 1988) and that potential threats in the workplace are overlooked. Logically, this approach also leads to one-sided recommendations to reduce stress, that is concentrating more on the individual employee than on changing the stressful condition.

The nature and traditions of psychology and occupational medicine

Traditionally, also, psychology and occupational medicine are biased towards the individual. Many psychology-oriented stress researchers are primarily interested in stress as a subjective and individual phenomenon. This may be a legacy of the strong tradition in psychology to focus on individual differences (i.e. differential psychology) and on individual counselling and therapy (i.e. clinical psychology and psychotherapy). In a similar vein Cooper and Cartwright (1994, p. 458) have stated that 'the professional "interventionists", the counsellors, physicians and clinicians are more comfortable with changing individuals than changing organizations'. Also, for most occupational physicians the core of their profession is to cure individual 'patients' (employees). Generally, occupational doctors feel more at ease in their expert medical role than in the role of organizational consultant or interventionist. It is beyond doubt that the question 'Why is it that employee A does develop stress complaints or reports sick, and why is it that employee B does not?' constitutes a valid and important research question in occupational health psychology and in occupational medicine. From the angle of prevention, however, a second question should be posed as well: 'Why is it that in certain occupations, in certain departments, and in certain companies, significantly more stress complaints are developed and that sickness absenteeism is significantly higher when compared to other occupations, departments and companies?'. Let us, for example, consider the occupation of city bus driver. Multimethodological, international data (e.g. Evans and Johansson, 1998) show that urban bus driving is a classic example of an occupation with high risks of occupational ill health. Pre-retirement work disablement is a major reason for leaving this occupation. Of course we could ask 'Why does bus driver C reach the official retirement age and why does bus driver D not?' However, we feel that the more focused question, at least from the point of view of ill-health prevention, is why it is that all over the world so many bus drivers develop serious stress-related health problems that cause them to leave this occupation involuntarily.

In this context, a warning seems appropriate against 'psychologism', that is, the explanation of societal events by individual psychic factors. The potential impact of more 'objective' or 'collective' factors in the work situation (box 1 in Figure 9.1; e.g. little control, a bad management style, too high demands) may then be – unjustly – lost sight of (see also Frese and Zapf, 1988).

Stressors may be inherent to the job

Stressors may be 'part of the job'. 'By definition', policemen and employees on psychiatric wards are confronted with violent people. 'By definition', nurses at oncological departments work with terminal cancer patients. Ambulance drivers and fire fighters will inevitably be confronted with seriously wounded or dead children. The point is that some work is stressful in itself, and that it may be not very realistic to reduce or eliminate all these risk factors. In those circumstances it makes sense to teach employees to cope with the necessary conditions of the job. The crucial issue here, of course, is to be able to distinguish between these necessary conditions and conditions that can be changed.

In this context we feel that one should bear in mind that jobs are not 'God-given'. They have been created and thus can be recreated (job redesign). In sum: to some extent the popularity of individual strategies to reduce stress may be explained by the fact that some stressors are intrinsic to the job. Training employees to cope with these demands is important, probably because it increases their feelings of control. These preventive measures should, however, not become a substitute for risk analysis and job redesign or an excuse for leaving the workplace as it is. The possibility to use adequate coping strategies depends on the level of control that an employee can exert over his or her work situation. For example, teaching city bus drivers how to provide service and how to deal with difficult passengers (which is an important task demand) is only realistic when the time schedule permits the drivers to use these skills (see also Table 9.1).

It is difficult to conduct methodologically 'sound' intervention and evaluation studies in a hectic organizational arena

Organizations are dynamic open systems and not laboratories. The difficulty of conducting methodologically 'sound' intervention and evaluation studies in a hectic organizational arena is often underestimated. Today, not only is the context of work rapidly changing, but also work itself. Work organizations keep transforming due to, among other factors, new production concepts (e.g. team-based work, lean production, telework), the flexibilization of work, the 24-hour economy, the increased utilization of information technology and the changing structure of the workforce. A related problem is that the major goal of most companies is not to facilitate 'sound scientific research', involving 'scientific outsiders' and detailed data collection on the scene. Organizational decision-makers may even regard research as a nuisance to the primary organizational processes. Against this background it may be well understood that the 'sound' evaluation studies that have been published are mostly of the 'individual type' (quadrants 3 and 4 in Figure 9.2; for an overview (Murphy, 1996). It is often easier, though not easy, to design an intervention study that compares an intervention group (usually receiving a kind of training course) and a (randomized) control group not (yet) receiving the treatment, than a study in which, after a risk assessment, piloting and thorough preparations, half of the company starts to work in autonomous teams and the other half not. In the section on methodological issues below, we will further discuss various methodological issues in stress intervention research.

The denominational segregation in stress research

A fifth cause may be found in the denominational segregation of stress research, with its relative neglect of studying the costs and benefits of stress prevention. Work and organizational psychologists concentrate primarily on 'soft' outcome variables (e.g. satisfaction, affect, moods and health complaints), mostly measured through questionnaires. Traditionally, it has been rather unusual for stress researchers to co-operate with economists in order to study the potential 'hard' outcome measures (e.g. productivity, company registered sickness absence rates, or accident rates) as well as the financial effects of interventions. To put it another way, a tradition of empirical insight into costs and benefits is absent in stress intervention research. It may therefore be well understood, as Cooper and colleagues (1996) have stated, that organizational decision-makers consider it 'easier and less disruptive to business to change the individual than to embark on any extensive and potentially expensive organizational development programme – the outcome of which may be uncertain' (p. 90).

Whereas work and organizational psychologists have often stated that an effective stress prevention programme may positively affect productivity and sickness absenteeism, until now they have not laid down a sufficiently strong empirical basis for this position. In a recent paper, Briner and Reynolds (1999) hold the view that this may be 'wishful thinking' even suggesting that: 'The misplaced optimism and moral sentiment that characterize the enthusiastic endorsement by most writers in the field of organizational interventions have, in fact, acted as a barrier to further development and elaboration of research and practice.' With Briner and Reynolds, we would emphasize the need for a more evidence-based strategy. In such a strategy and in order to increase the impact of stress prevention on what organizational decision-makers really consider important (e.g. quality of product and services, organizational flexibility, continuity, absenteeism, labour market considerations and improved productivity), stress researchers should definitely extend their focus by also including 'hard' outcome variables (e.g. productivity and sickness absenteeism). Therefore more co-operation with other disciplines, such as economists and ergonomists, is desirable.

Further methodological issues in stress intervention research

Above we have argued that it is difficult to conduct methodologically 'sound' intervention research in today's organizations, and that this is one of the reasons why organization-level stress interventions are scarce. Now we will address some methodological issues in stress research in more detail, starting with (a) a commentary on the common lack of a proper risk analysis, and with a discussion of the 'true experiment' and some validity questions (b–c). Next we move on to a discussion (d) of randomization problems in psychosocial intervention research (e) of aetiological versus feasibility studies, and finally (f) of four common shortcomings in stress intervention research.

(a) The lack of a proper analysis

An intriguing methodological issue is the lack of a proper problem analysis, in terms of the identification of risk factors and risk groups, in many prevention programmes. We feel that this is both curious and unacceptable. Already in 1975, Caplan and colleagues (cited in Newman and Beehr, 1979, p. 21) concluded that: 'A thorough diagnosis of the problem is critical. The types of job stressors must be identified as well as the extent of their relationship to the state of employee health.' In a similar vein Cooper, Liukkonen and Cartwright (1996) comment: 'An organization needs to know its starting point, in order to assess the benefits derived' (p. 71). We would argue that without knowledge with respect to causes, prevention is impossible. Often, problems that can be understood can also be solved. In other words, in a stress reduction project first the two basic questions with respect to Figure 9.1 should be answered: i.e. 'What's in every box?', and 'How are these boxes interrelated?'.

(b) Three intervention study designs, internal and theoretical validity

According to Campbell and Stanley (1966), intervention study designs may be categorized as truly experimental, quasi-experimental, and non-experimental.

These designs differ as to 'who gets measured and when'. The true experiment includes randomly assigned experimental and control groups to reduce confounding and selection bias. The quasi-experiment uses a non-randomly assigned experimental group and control group, and the non-experimental design has an experimental group only. The timing of measures can include pre-test and post-test, post-test only, and time-series (Cook and Campbell, 1979). A common misperception is that these three research strategies should be arrayed hierarchically, the true experiments ranking highest and the non-experimental cases receiving the lowest ranking (Yin, 1994). True experiments offer the best potential for making causal inferences. We should bear in mind, however, that true experiments do not guarantee that causal inferences can reasonably be made, or that causal relations between variables reflect causal relations between the higher order constructs that they are supposed to operationalize. The latter point is a matter of theoretical or construct validity. We will first turn to the first point. This is a matter of internal validity, i.e. are the effects caused by the independent factors? Causal inferences are internally valid only when the observed change or difference can be attributed confidently to a specific variable that has been identified or isolated by the investigator (Neale and Liebert, 1986). With how much confidence such inferences can be made depends largely on a critical examination of possible threats to internal validity, i.e. potential third variables that can invalidate these relations. Such threats are: maturation, testing, instrument decay, statistical regression, history, selection, mortality, diffusion of treatments, and competition by people receiving no treatment (Neale and Liebert, 1986; Beehr and O'Hara, 1987).

Not only internal validity counts but also the ability to generalize effects: (a) theoretical validity, and (b) external validity. Theoretical (or construct) validity involves the ability to generalize the independent and dependent variables to higher-order constructs, whereas external validity entails the extent to which the results can be generalized to and across populations of persons, settings and times (Beehr and O'Hara, 1987). Without external validity one does not know whether similar intervention programmes would be equally effective if implemented again. Internal validity is a conditio sine qua non but again no guarantee for these kinds of validity. 'Both external validity and construct validity are a matter of replication of the experiment, especially replication at many times and with a variety of situations and people' (Beehr and O'Hara, 1987, p. 90).

There is a related question with respect to construct validity: i.e. 'Are most of the person-targeted treatments treating stress'? Beehr and O'Hara (1987) first raised this question. They argue: 'Another consequence of the overwhelming use of person-targeted treatments in virtually all published reports of stress treatments in the workplace is that there is no evidence that the treatment is treating stress at all' (p. 107). They notice that certain individual responses such as adrenaline excretion, cholesterol, blood pressure are regarded as stress effects by journal editors and the mass media. Any programme that treats these variables is therefore labelled a stress treatment, and this label as well is accepted by the editors of scientific journals. We agree with their argument: 'The point is that there are no stress treatments unless there is some stressor known to have caused or at least been related to these responses, and the literature on stress-management programmes is virtually devoid of any such evidence' (Beehr and O'Hara, 1987, p. 107).

(c) The true experiment in a real organizational context?

As emphasized, it is extremely difficult and often impossible to transfer the true experimental design to the practical reality of today's organizations. The problem for researchers is that interventions always take place in context, and that this context is not under control of scientists. In this context people may trust or distrust each other and have different or shared interests. They are not passive study objects, instead they are active organizers of their own (working) situation based on their interests, preferences and attitudes.

In such a context there may well be competing demands between scientific aims and practical aims. Whereas scientific aims emphasize extensive and detailed analyses of stressors and strains (e.g. triangulation), employees and/or management often want changes. Often, in intervention projects, the question 'When do we know enough?' will be answered differently by scientists and by employees and organizational decision-makers. In practice, further data gathering and data analysis may even inhibit organizational improvement, letting 'the momentum fade away'. A further example of 'competing demands' between academic research and organizational practice is that longitudinal data collection, to be preferred from a research angle, may inhibit further participation and leave intervention projects with a biased sample (Lourijsen, Houtman, Kompier and Gründemann, 1999; Kalimo and Toppinen, 1999).

There is another possible complication with the true experiment, which we refer to as 'reductionism'. Until now little attention in stress intervention research literature has been given to this phenomenon. The core of the true experiment is to isolate stimulus and response (while controlling for third 'contextual' factors), in order to make causal S–R inferences. Two common types of reductionism in stress (intervention) research deserve attention. The first is the belief that a single causal relationship can be modified by an intervention (Goldenhar and Schulte, 1994). This again is a matter of construct or theoretical validity. Neither stressors (box 1 in Figure 9.1), nor stress effects or consequences such as health deterioration or chronic fatigue (boxes 3 and 4) are discrete variables. 'In reality, these causal phenomena are complex, multistep causal processes with wide ranges of outcomes and may extend over time' (Goldenhar and Schulte, 1994, p. 770). By trying to isolate stimulus and response in order to increase internal validity one usually gives in, i.e. loses, theoretical validity. The second issue is that research that focuses exclusively on outcomes limits the understanding of a phenomenon in field settings. We feel that traditional stimulus–response schemata are often incompatible with an understanding of the dynamics and effectiveness of stress reduction programmes in field settings. Such 'technical' cause–effect reasoning tends to underestimate the crucial role of the introduction and implementation of such programmes, that is of contextual and 'process variables'. In the same vein, Goldenhar and Schulte (1994) conclude that the complexity of phenomena in occupational health intervention studies also means that intervention researchers should focus more on the process and the milieu of an intervention and not only on the outcomes. With respect to work redesign, Parker and Wall (1997), while reviewing key issues for future research, arrive at a comparable conclusion: 'To this point we suggest the greater use of qualitative approaches to allow a better understanding of the complex, and often highly political, dynamics that are involved in work redesign. We also advocate the wider reporting of "process issues" (in most published work design studies, the focus is on outcomes rather than process)' (p. 137). In a similar vein, Ovretweit (1998), while discussing the evaluations of health interventions, concludes: 'Traditional experimental evaluation design is not well suited to investigating social systems or the complex way in which interventions work with subjects or their environment' (p. 99).

(d) Problems with randomization

Before we discuss the problems of randomization in psychosocial intervention research, we should try to understand why randomization was introduced and what purposes it serves. In aetiological intervention research the purpose of a study is to assess the effect of an intervention with regard to one or more strictly defined endpoints. It is not enough to give 100 persons with low back pain a treatment and see if they get better or not. Our problem is that an unknown proportion of the 100 persons would have got better without the treatment. In order to follow the natural course of events without treatment we need a control group. The effect of the treatment is per definition the difference between the natural course of events (represented by the control group) and the course of events in the treatment group. In order to make the comparison as valid as possible, the two groups should be as identical as possible. The solution to this problem is usually randomization, where a random procedure decides the distribution of the target group into the intervention and control group. By randomizing we try to tackle two important problems at the same time: confounding and selection bias. We have confounding if known or unknown risk factors for the endpoint are unevenly distributed in the two groups, and we have selection bias if those in the intervention group are more motivated or in other ways are 'biased' compared with the control group.

Randomization has been a unique solution to the problems of confounding and selection bias in biomedicine and psychology since it was introduced more than 100 years ago. The method does, however, have a number of limitations of which we shall mention a few of special relevance for psychosocial work environment research.

First of all, in biomedicine and psychology the unit of randomization has always been the individual. A sufficient number of persons (or animals) have been randomized to the treatment and control group(s), and the course of events for these individuals has been compared. In work environment intervention studies the unit of intervention is often a work site or a department. This is almost always the case when the intervention is organizational or includes the interpersonal relations of a whole group of employees. In such situations the number of potential intervention and control departments is usually very limited and randomization makes little sense. Often it will be wiser to select the intervention and control departments in pairs so that each intervention department has a similar control work site.

The procedure of selecting work sites can in itself be extremely problematic in practice. This has to do with a basic dilemma in psychosocial intervention research. If the intervention is participatory and based on the active involvement, enthusiasm, and activity of the workers, then it might be impossible to 'select' the intervention and control groups in an optimal way. It simply makes very little sense to offer the workers participation in a 'bottom-up' participatory psychosocial intervention study, and then afterwards decide that they should belong to a 'control' group without involvement and enthusiasm. And it makes equally little sense to tell people that they have been selected to be enthusiastic and involved. It is simply not possible to treat workers and supervisors as guinea pigs, and the option of a 'placebo psychosocial intervention' is out of the question for ethical and many other reasons.

The fact that the model of the randomized control trial usually cannot be applied in psychosocial work environment intervention studies does not imply that we should neglect the problems of confounding and selection bias. On the contrary, these and other methodological problems should be dealt with as effectively as possible, but it should always be kept in mind that methods are tools, not goals in themselves. There are many ways in which we can elucidate selection processes and control for confounding factors, and these methods should be used here as elsewhere (e.g. Hernberg, 1992).

(e) Aetiological versus feasibility studies

Intervention studies are usually discussed as aetiological studies: if the change in the intervention group is significantly larger than in the control group this is interpreted as an effect of the intervention (the cause). It should, however, be pointed out that the intervention study is an excellent tool for studying an equally important problem: the problem of prevention effectiveness or feasibility.

This point can be illustrated by an imaginary example. Suppose that we initiate an intervention study in order to test the following hypothesis: 'Improvement of communication skills can reduce the level of interpersonal conflicts among workers who work in groups.' We test this hypothesis by performing an intervention study in which a number of work groups participate in a course in assertive communication skills while the control groups do not. (They are on a waiting list and will participate in the course some time in the future.) In most studies of this type the level of interpersonal conflicts is measured before the course and some time after the course in both the intervention and control groups. If the effect is as expected, it is concluded that the hypothesis was confirmed and if not, the opposite is concluded. This may not be a very prudent way to conduct an intervention study. Table 9.1 illustrates the problem.

The first three questions (1a–1c) have to do with the quality of the course in communication skills. It is surprising that most studies in the literature suppose that the intended effect of a course (or of any other intervention) is the same as the actual effect. Many courses are quite short, and the participants are rarely learning how to use the principles of the course in real-life situations. Nevertheless, many intervention researchers are surprisingly uninterested in elucidating these important questions.

The next three questions (2a–2c) have to do with the situation after the participants return to their former work. Even in the ideal situation where all the colleagues have attended the course together, there may be many

Table 9.1: Feasibility and aetiological questions in an intervention study on communication skills and interpersonal conflicts.

Research question: Can improvement of communication skills reduce the level of interpersonal conflicts among workers who work in groups?

Specific research questions:

1. The course: Did the participants acquire the skills?
   a. How many attended the course?
   b. How much did the participants learn during the course?
   c. Were the participants trained in practising the new skills?

2. Implementation after the course: Did the participants use their new skills?
   a. Was it possible for the participants to use their new skills?
   b. Were the participants able and willing to use their new skills?
   c. For how long and to what extend did they practise the new skills?

3. The effects: Did the use of new communication skills influence the occurrence of interpersonal conflicts?
   a. Did the new skills reduce conflicts between those who already had conflicts?
   b. Did the new skills prevent new interpersonal conflicts among those who received training?
   c. Did the new skills prevent conflicts between course participants and other colleagues?

barriers blocking the implementation of the new skills. The time pressure may be too great, or some colleagues may lack the courage or energy it takes to practise assertive communication. They may also feel that supervisors or colleagues do not support the use of the new skills.

All the first six questions have to do with feasibility (or prevention effectiveness) and it goes without saying that the important question of effect has no meaning at all, if these six questions are not answered satisfactorily.

The last three questions (3a–3c) have to do with aetiology. Do the new communication skills reduce the occurrence of interpersonal conflicts as hypothesized? Again, the question is more complicated than often assumed. In most studies the average level of e.g. conflicts, stress, absence from work, burnout or low back pain is measured before and after the intervention. This is usually much too superficial. First of all, preventing new cases of conflicts from occurring is not the same as reducing the level of conflict between those who already have one. The same goes for stress, burnout or medical diseases: changing the risk of becoming ill should not be confused with treating those who are already ill. Yet this is exactly what researchers do in most intervention studies.

By distinguishing between the questions of feasibility and aetiology we separate the two central questions in intervention research: Did the patient take the (intended) pill? And: Did the pill have (the intended) effect? Needless to say, it does not help to take the pill if it has no effect (aetiology). It is, however, equally useless that the pill has effect if it is not taken (feasibility). If an intervention study shows no effect of the intervention (which is not uncommon), then it is of paramount importance to be able to separate the questions of feasibility and aetiology.

(f) Common errors in intervention research

In many reviews of intervention research a number of common errors or shortcomings are pointed out (Newman and Beehr, 1979; Goldenhar and Schulte, 1994; Skov and Kristensen, 1996; Van der Hek and Plomp, 1997; Kristensen and Borritz, 1998). The most common shortcoming is the lack of differentiation between aetiology and feasibility as discussed above. There are four other shortcomings that are so common that they deserve to be mentioned briefly here.

- *Floor or ceiling effects*. In many studies the level of the effect measure (such as burnout, depression or absence from work) is so low in both intervention and control groups that any effect of the intervention would seem very unlikely from the very beginning. In fact, some interventions resemble smoking cessation courses for non-smokers. For example, a number of burnout intervention studies are initiated because the studied groups (such as nurses or social workers) are expected to have high levels of burnout according to the literature. In many cases these interventions have been planned before the low level of burnout was known by the researchers. In some cases the pre-intervention level of burnout was not calculated until the end of the intervention period (Kristensen and Borritz, 1998).
- *Lack of differentiation*. In most intervention studies all participants are treated as if they were identical. This is perhaps in accordance with the statistical model of 'random selection of interchangeable units', but from a theoretical point of view it is usually very interesting to study differential effects of an intervention. First of all, it would be of interest to differentiate between persons with high, middle and low levels of the variable under study (such as burnout). This would deal effectively with the dilution effect due to the participants with low burnout. Second, it would also be interesting to study differential effects with regard to personality types, gender, age, education, et cetera. Such differential analyses (sub-group analyses) should be performed much more often in future intervention studies.

- *Lack of distinction between statistical and practical/clinical significance.* The effect of an intervention is assessed with the help of statistical significance testing in the majority of intervention studies. We will not go into details with regard to statistical testing here, but only point to the obvious problem that statistical significance is a result of two factors: the size of the samples under study and the size of the differences found. The larger the population, the less it takes to reach statistical significance. This means that large differences are ignored in studies with small group sizes while very small differences are labelled as 'significant' in large studies. In those cases where the differences are found to be significant, the distinction between statistical and practical/clinical significance is rarely discussed. How big should a difference on a burnout or stress scale be in order to be noticeable and meaningful for the persons involved? How big should the differences be in order to make a difference with regard to prognosis? Researchers ought to address these and similar issues much more often.
- *Too short follow-up.* Also the lack of adequate follow-up time has been noticed in many reviews. In the review by Kristensen and Borritz (1998) it was found that several of the burnout intervention studies measured burnout just before a course and then again on the last day of the course. Needless to say, this is insufficient if one wants to assess the effect of an intervention. From an ideal point of view the best design would be to measure the endpoint variables several times after the intervention period in order to be able to distinguish between short- and long-term effects. In cancer research the golden rule is five years of follow-up after treatment, and a similar time span would clearly be relevant in much psychosocial intervention research.

Future

Looking back

Until now we have argued that while true experiments offer the best potential for causal inferences, they do not guarantee that causal inferences can reasonably be made, or that associations between variables reflect causal relations between the higher order constructs that they are supposed to operationalize. We have also argued that in stress intervention research it often is very difficult to transfer the true experimental paradigm to the practical reality of today's organizations because interventions take place in a context – which is not under control of scientists – and because the 'people under study' (employees, supervisors, managers) are not passive study objects, but active organizers of their own (working) situation.

In addition, we have stated that traditional 'technical cause–effect' reasoning does not sufficiently take into account the role of these contextual variables and process variables, such as the introduction and implementation of interventions.

Furthermore, we have discussed the need for a proper analysis of risk factors and risk groups, and discussed in more detail some problems with randomization in psychosocial intervention research. We have also emphasized the crucial distinction between aetiology ('Did the pill have the – desired – effect?') and feasibility ('Did the patient take the – intended – pill?'). Finally, we have discussed four other shortcomings in psychosocial intervention research: floor or ceiling effects, lack of differentiation, lack of distinction between statistical and practical/clinical significance, and a too short follow-up.

Looking ahead: recommendations and multiple case study approach

Recommendations for future stress intervention research

Several recommendations for future stress intervention research follow from our discussion of the past and present of stress intervention research and the methodological issues that were discussed above. We would suggest that future stress intervention studies should include some of the following:

1. *Theories*. Intervention studies should be based on explicit theories. Against the background of our discussion of internal and theoretical validity, this means that interventions should theoretically and logically 'fit in' with the problems that have been identified ('the key should fit in the keyhole').
2. *Diagnosis*. An adequate diagnosis, identifying risk factors and risk groups ('lack of differentiation'), is a conditio sine qua non for each preventive programme. If the effect measure (e.g. burnout, sickness absenteeism) is low, we do not need an intervention ('floor and ceiling effects'). If certain groups at risk can be identified, sub-group or differential analyses should be performed for these groups.
3. *'Soft' and 'hard' outcomes*. Stress researchers should not only address 'soft' outcome variables (e.g. motivation, satisfaction, health complaints), but extend their focus to also include 'hard' outcome variables (e.g. productivity and sickness absenteeism). In order to increase the impact of stress prevention in the workplace, more emphasis should be placed on such factors as the quality of products and services, organizational flexibility, continuity,

absenteeism, productivity, labour market facets, and improved competitiveness; and for there to be a multi-disciplinary approach rather than
the traditional mono-disciplinary one (e.g. co-operation with economists and ergonomists). Endpoints should be clear from the start of the
study and should be measured in a valid way.
4. *Aetiology and feasibility*. Two central questions in intervention
   research should be distinguished, i.e. the question of aetiology and of
   feasibility. The distinction between aetiology and prevention effectiveness should be kept clear from the beginning.
5. *Statistical significance and practical relevance*. Researchers should
   remember that not everything that is significant is relevant and, vice
   versa, that not everything that is non-significant is irrelevant (for
   example, due to small numbers).
6. *Follow-up*. The follow-up time in intervention projects should be
   adequate. Ideally, the best design would be to measure both short-term
   and long-term endpoint variables, at the same time studying selective
   attrition. The time frame should be theoretically derived – usually
   longer than today.

In addition to these recommendations, in order to answer the questions:
'How well, why and when do stress prevention programmes work?' we do
need more examples of (good) preventive practice. Therefore, we would
like to advocate an alternative research strategy in stress intervention
research, i.e. a multiple case study approach.

Multiple case study approach

This alternative approach, which is not a replacement of but an addition to
the true experimental paradigm, is centred around 'natural experiments'
in which data are collected before and after some relevant intervention in
the workplace.

The case study has long been stereotyped as a weak sibling among
social science methods (Yin, 1994, p. xiii). However, a multiple case study
approach is an adequate research strategy when 'how' or 'why' questions
are being posed, when the investigator has little control over events, and
when the focus is on a contemporary phenomenon within some real-life
context (Yin, 1994). In the context of this chapter, occupational stress is
the 'contemporary phenomenon', and the organization is the 'real-life
context'.

As Yin (1994) points out, multiple cases should be considered
as multiple experiments or multiple surveys (i.e., follow a replication
logic), instead of as multiple respondents in a survey. Accordingly, the
method of generalization is 'analytic generalization' and not 'statistical

generalization'. In statistical generalization, inferences are made about a population on the basis of empirical data collected about a sample. Cases, however, do not represent a 'sample': they are generalizable to theoretical propositions and not to populations. Such a theoretical proposition, for example, is that increasing job control will reduce health complaints, or that the introduction of team-based work will increase the motivation for learning new behaviour patterns and productivity. This research strategy has been characterized as 'plausible rival hypotheses' by Campbell (1994, p. ix): 'The core of the scientific method is not experimentation per se but the strategy connoted by the phrase plausible rival hypotheses'. This research strategy is quite similar to principles used in criminal investigations (Yin, 1994). As for the court, in stress intervention research 'full proof' of (causal) relationships, e.g. the relationship between work redesign and sickness absenteeism, is hard to give. The essence of 'plausible rival hypotheses' is that a researcher, similar to a prosecutor or a lawyer, systematically brings up arguments and draws conclusions with respect to plausibility by systematically looking for converging (e.g. triangulation) and diverging evidence (i.e. competing causes or artefacts that may otherwise account for the observed outcomes).

In multiple case studies, research data can be treated cumulatively. We feel that such multiple case studies may constitute an adequate research strategy for several reasons. They may provide an alternative to the 'true experimental approach' that was characterized earlier, and also to the great majority of cross-sectional questionnaire studies with very limited evidence regarding the causal role of work characteristics for health outcomes (Kasl, 1978). Also, they may function as alternatives to many of the post hoc individual-biased studies that now dominate the literature on work stress interventions. Furthermore, they may provide useful information with respect to the role of contextual and process factors in stress prevention programmes. Finally, case studies are important because of 'the power of the good example', the adage being that 'good examples do follow'.

Empirical research in the proposed direction is still somewhat scarce, but in the last decade progress has been made (e.g. Karasek, 1992; Burke, 1993; Kompier et al., 1998; Kompier and Cooper, 1999; Kompier, Aust, Van den Berg and Siegrist, 2000). To illustrate this multiple case approach, we will consider the latter study on stress prevention in bus drivers in a little more detail. Few other contemporary professions are as stressful as urban bus driving. However, in the literature on occupational stress of bus drivers, there is a remarkable distinction between the impressive number of studies that demonstrate adverse health effects of the bus drivers' occupation and the small amount of documented prevention and intervention projects in bus companies. Therefore, in the latter study (Kompier et al., 2000) the research aim was to select, compare and analyse interventions and preventive actions from international bus companies in order to decrease bus drivers' occupational stress and sickness absenteeism. Through networking, international surveys and literature study thirteen 'natural experiments' were identified with an acceptable research design rating. Interventions were both work- and person-directed. Principles of worker participation were often followed. The variety in intervention programmes, outcome measures, case evaluations and methodological flaws made it difficult to present a general picture of programme effectiveness. The study suggests that stress prevention that combines adequate interventions and proper implementation may be beneficial to both the employee and the company.

Postscript

From the current overview of the past and present of research into occupational stress we conclude that since the pioneering efforts of Newman and Beehr (1979) a lot of progress has been made. Work and organizational psychologists have indeed taken up the gauntlet. Both theorists and practitioners have been very active in this area. Not all developments are positive, however. Rather independently of scientific progress, stress management has become a commercial market with doubtful quality. Although there is more theoretically and methodologically sound intervention research than twenty years ago, this research still has a 'post hoc bias to the individual'. There still are too few work and organization directed intervention studies, and there are various methodological pitfalls that should be addressed. In sum: it is our opinion that during the last decades serious progress has been made in the field of work, stress and health ('occupational health psychology'). We feel that in the years to come the main challenge of occupational health psychology is to transform this impressive existing body of knowledge on 'stress and health' into prevention. Hopefully the presented recommendations will stimulate this development.

References

Beehr TA, O'Hara L (1987) Methodological designs for the evaluation of occupational stress interventions. In: Kasl SV, Cooper CL (eds) Research Methods in Stress and Health Psychology, pp. 79–112, Chichester: Wiley.
Briner RB, Reynolds S (1999) The costs, benefits, and limitations of organizational level stress interventions. Journal of Organizational Behavior 20(5):647–64.
Burke RJ (1993) Organizational-level interventions to reduce occupational stressors. Work and Stress 7(1): 77–87.

Campbell DT (1994) Foreword. In: Yin RK, Case Study Research. Design and Methods, 2nd edn. Sage: Thousand Oaks, CA.

Campbell DT, Stanley JC (1966) Experimental and Quasi-experimental Designs for Research. Chicago, IL: Rand-McNally.

Cook TD, Campbell DT (1979) Quasi-experimentation: Design and Analysis Issues for Field Settings. Chicago, IL: Rand McNally.

Cooper CL (ed.) (1998) Theories of Organizational Stress. Oxford: Oxford University Press.

Cooper CL, Payne R (eds) (1988) Causes, Coping and Consequences of Stress at Work. Chichester: Wiley.

Cooper CL, Cartwright S (1994) Healthy mind; healthy organization. A proactive approach to occupational stress. Human Relations 47(4): 455–71.

Cooper CL, Liukkonen P, Cartwright S (1996) Stress prevention in the workplace: Assessing the costs and benefits to organisations. Dublin: European Foundation for the Improvement of Living and Working Conditions.

Cox T (1993) Stress research and stress management: putting theory to work. (HSE contract research report, no. 61/1993) Nottingham: Centre for Organizational Health and Development, University of Nottingham.

DeFrank RS, Cooper CL (1987) Worksite stress management interventions: their effectiveness and conceptualization. Journal of Managerial Psychology 2: 4–10.

Evans GW, Johansson G (1998) Urban bus driving: an international arena for the study of occupational health psychology. Journal of Occupational Health Psychology 3(2): 99–108.

Frese M, Zapf D (1988) Methodological issues in the study of work stress: objective versus subjective measurement of work stress and the question of longitudinal studies. In: Cooper CL, Payne R (eds) Causes, Coping and Consequences of Stress at Work, pp. 375-411. New York: Wiley.

Frone MR, Russell M, Cooper ML (1992) Antecedents and outcomes of work-family conflict: testing a model of the work-family interface. Journal of Applied Psychology 77: 65–78.

Goldenhar LM, Schulte PA (1994) Intervention research in occupational health and safety. Journal of Occupational Medicine 36(7): 763–75.

Griffiths A, Cox T, Barlow C (1996) Employers' responsibilities for the assessment and control of work-related stress: a European perspective. Health and Hygiene 17: 62–70.

Hernberg S (1992) Introduction to Occupational Epidemiology. Chelsea, MI: Lewis Publishers.

Ivancevich JM, Matteson MT, Freedman SM, Phillips JS (1990) Worksite stress management interventions. American Psychologist 45: 252–61.

Kahn RL, Byosiere P (1992) Stress in organizations. In: Dunnette MD, Hough LM (eds) Handbook of Industrial and Organizational Psychology, 2nd edn, vol. 3, pp. 571–650. Palo Alto: CA: Consulting Psychologists Press.

Kalimo R, Toppinen S (1999) Finland: Organizational well-being. Ten years of research and development in a forest industry corporation. In: Kompier MAJ, Cooper CL (eds) Preventing Stress, Improving Productivity. European Case Studies in the Workplace, pp. 52–85. London: Routledge.

Karasek R (1992) Stress prevention through work reorganization: A summary of 19 international case studies. In: ILO (Di Martino V, ed.) Conditions of work digest. Preventing Stress at Work 11(2): 23–41.

Karasek RA, Theorell T (1990) Healthy Work. Stress, Productivity and the Reconstruction of Working Life. New York: Basic Books.

Kasl SV (1978) Epidemiological contributions to the study of work stress. In: Cooper CL, Payne R (eds) Stress at Work. Chichester: Wiley.

Kasl SV (1987) Methodologies in stress and health: Past difficulties, present dilemmas, future directions. In: Kasl SV, Cooper CL (eds) Stress and Health: Issues in Research Methodology. Chichester: Wiley.

Kasl SV, Cooper CL (eds) (1987) Research methods in Stress and Health Psychology. Chichester: Wiley.

Kompier MAJ, Cooper CL (eds) (1999) Preventing Stress, Improving Productivity. European Case Studies in the Workplace. London: Routledge.

Kompier MAJ, De Gier E, Smulders P, Draaisma D (1994) Regulations, policies and practices concerning work stress in five European countries. Work and Stress 8(4): 296–318.

Kompier MAJ, Geurts SAE, Gründemann RWM, Vink P, Smulders PGW (1998) Cases in stress prevention: the success of a participative and stepwise approach. Stress Medicine 14: 155–68.

Kompier MAJ, Aust B, Van den Berg A, Siegrist J (2000) Stress prevention in bus drivers: evaluation of thirteen natural experiments. Journal of Occupational Health Psychology 5(1): 11–31.

Kristensen TS, Borritz M (1998) Forebyggelse af udbraendthed (Prevention of burnout). Copenhagen: The Working Environment Fund.

Levi L (1984) Work, stress and health. Scandinavian Journal of Work, Environment and Health 10: 495–500.

Lourijsen E, Houtman I, Kompier M, Gründemann R (1999) The Netherlands: a hospital, 'Healthy working for health'. In: Kompier MAJ, Cooper CL (eds) Preventing Stress, Improving Productivity. European Case Studies in the Workplace, pp. 86–120. London: Routledge.

McLeroy KR, Bibeau D, Steckler A, Glanz K (1988) An ecological perspective on health promotion programs. Health Education Quarterly 15: 351–77.

Murphy LR (1984) Occupational stress management: A review and appraisal. Journal of Occupational Psychology 57: 1–15.

Murphy LR (1986) A review of organizational stress management research. Journal of Organizational Behavior Management 8: 215–27.

Murphy LR (1996) Stress management in work settings: A critical review of the health effects. American Journal of Health Promotion 11(2): 112–35.

Murphy LR, Hurrell JJ, Sauter S, Keita G (eds) (1995) Job Stress Interventions. Washington, DC: American Psychological Association.

Neale JM, Liebert RM (1986) Science and Behavior. An Introduction to Methods of Research, 3rd edn. Englewood Cliffs, NJ: Prentice-Hall.

Newman JE, Beehr TA (1979) Personal and organizational strategies for handling job stress: a review of research and opinion. Personnel Psychology 32: 1–43.

Ovretweit J (1998) Evaluating Health Interventions. Buckingham-Philadelphia, PA: Open University Press.

Parker S, Wall T (1997) Job and Work Design. Organizing Work to Promote Well-being and Effectiveness, Thousand Oaks, CA: Sage.

Quick JC, Murphy LR, Hurrell JJ (eds) (1992) Stress and Well-being at Work, Washington: American Psychological Association.

Quick JC, Quick J, Nelson DL, Hurrell JJ (1997) Preventive Stress Management in Organizations. Washington, DC: American Psychological Association.

Reynolds S, Briner R (1994) Stress management at work: with whom, for whom and to what ends? British Journal of Guidance and Counselling 22(1): 75–89.

Skov T, Kristensen TS (1996) Etiologic and prevention effectiveness intervention studies in occupational health. American Journal of Industrial Medicine 29: 378–81.

Spector PE (1992) A consideration of the validity and meaning of self-report measures of job conditions. In: Cooper CL, Robertson IT (eds) International Review of Industrial and Organizational Psychology. Chichester: Wiley.

Van der Hek H, Plomp HN (1997) Occupational stress management programmes: a practical overview of published effect studies. Journal of Occupational Medicine 47(3): 133–41.

Yin RK (1994) Case Study Research. Design and Methods, 2nd edn. Thousand Oaks, CA: Sage.

Index